DEGRADATION OF POLY (LACTIDE) - BASED BIODEGRADABLE MATERIALS

NOTICE TO THE READER

The Publisher has taken reasonable care in the preparation of this book, but makes no expressed or implied warranty of any kind and assumes no responsibility for any errors or omissions. No liability is assumed for incidental or consequential damages in connection with or arising out of information contained in this book. The Publisher shall not be liable for any special, consequential, or exemplary damages resulting, in whole or in part, from the readers' use of, or reliance upon, this material.

Independent verification should be sought for any data, advice or recommendations contained in this book. In addition, no responsibility is assumed by the publisher for any injury and/or damage to persons or property arising from any methods, products, instructions, ideas or otherwise contained in this publication.

This publication is designed to provide accurate and authoritative information with regard to the subject matter covered herein. It is sold with the clear understanding that the Publisher is not engaged in rendering legal or any other professional services. If legal or any other expert assistance is required, the services of a competent person should be sought. FROM A DECLARATION OF PARTICIPANTS JOINTLY ADOPTED BY A COMMITTEE OF THE AMERICAN BAR ASSOCIATION AND A COMMITTEE OF PUBLISHERS.

LIBRARY OF CONGRESS CATALOGING-IN-PUBLICATION DATA

Tsuji, Hideto.
  Degradation of poly (lactide)- based biodegradable materials / Hideto Tsuji.
    p. ; cm.
  Includes index.
  ISBN 978-1-60456-502-7 (softcover)
  1. Biodegradable plastics. 2. Polymers--Biodegradation. I. Title.
  [DNLM: 1. Polyesters--metabolism. QD 381 T882d 2008]
  TP1180.B55T78 2008
  620.1'92323--dc22

                              2008007301

Published by Nova Science Publishers, Inc. ✢ New York

CONTENTS

PREFACE

Poly(lactide), i.e., poly(lactic acid) (PLA) is a biodegradable polyester produced from renewable resources. Due to its high mechanical performance and very low toxicity, PLA is utilized for biomedical, pharmaceutical, ecological, industrial, and commodity applications. PLA is susceptible to various types of degradation, such as hydrolytic degradation in the human body and the environment, biodegradation and photodegradation in the environment, and thermal degradation during processing. In biomedical and pharmaceutical applications, the hydrolytic degradation mechanism and rate are crucial factors for determining its performance. In contrast, in most of the environmental, industrial, and commodity applications, an accurate adjustment of the degradation rate as accurate as in biomedical and pharmaceutical applications is not required, or a degradation during use can cause a reduction in material performance. However, for recycling PLA to its monomers, information on the mechanism and rate of hydrolytic degradation and thermal degradation at elevated temperatures is indispensable.

This book deals with the hydrolytic degradation, biodegradation, thermal degradation, and photodegradation of PLA-based materials with different material factors and degradation conditions or environments.

Chapter 1

INTRODUCTION

Poly(lactide), i.e., poly(lactic acid) (PLA) is biodegradable in the human body as well as in the environment, compostable, producible from renewable resources such as starch, and has very low toxicity to the human body and the environment [1-12]. Because of these reasons, PLA has been intensively studied for more than forty years in terms of scientific interest and a wide variety of applications. The schematic representation of synthesis, recycling, and biodegradation of PLA is shown in Figure 1. Also, PLA has high mechanical performance compared to that of representative commercial polymers such as polystyrene and poly(ethylene terephthalate) (PET). Because of its biodegradability and very low toxicity, PLA is utilized as biomedical material for tissue regeneration and matrices for drug delivery systems, as well as alternatives for commercial polymers. Recently, PLLA composite materials are used as automobile parts and the casing for personal computers and mobile phones.

In biomedical and pharmaceutical applications, in addition to the mechanical performance, the hydrolytic degradation mechanism and rate are crucial factors for determining its performance. However, in most of the environmental, industrial, and commodity applications, an adjustment of the degradation rate as accurate as in biomedical and pharmaceutical applications is not required, or a degradation during use can cause a reduction in material performance. PLA can be readily hydrolytically degraded to lactic acid and thermally depolymerized to lactide (the dimer of lactic acid). Therefore, these reactions can be utilized for recycling of PLA. In these cases, information on the mechanism and rate of hydrolytic degradation and thermal degradation at elevated temperatures is indispensable for attaining a high monomer yield within a short period. Recently, the price of corn, which is a raw starch material, has been soaring, due to the rapid increase in ethanol production from corn (Figure 1), mainly in the USA.

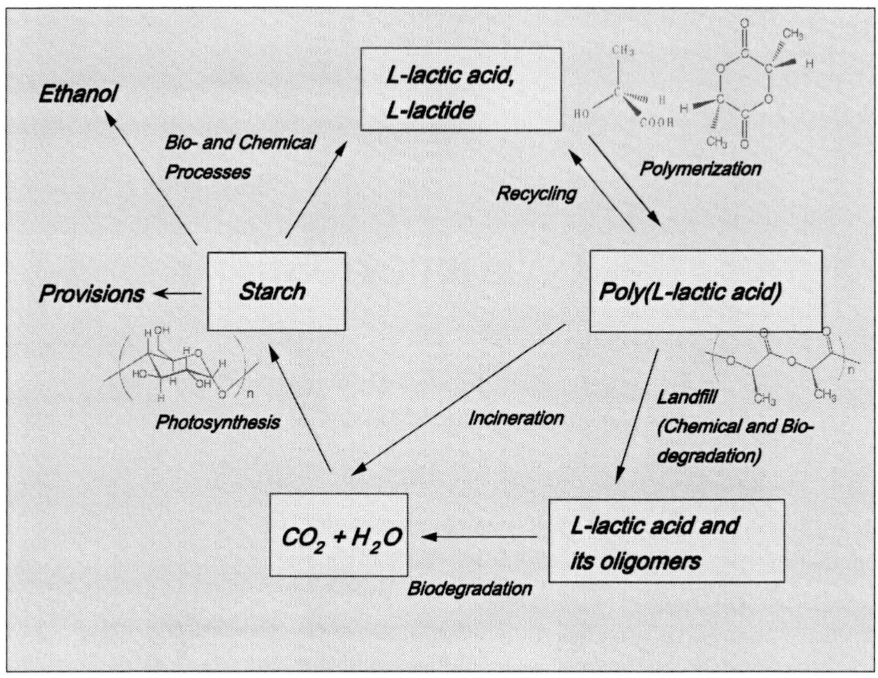

Figure 1. Synthesis, recycling, and degradation of poly(L-lactic acid) (PLLA).

This threatens to elevate the production cost of PLA. This review deals with various types of degradation of PLA-based materials having different molecular and highly ordered structures, fillers, material shape, and surface structure, and summarizes the effects of these material factors and degradation conditions or environments on the degradation mechanism and rate. The degradation of PLA-based materials reviewed in this article includes hydrolytic degradation, biodegradation, thermal degradation, and photodegradation. However, before discussing various types of degradation, the synthesis and applications of PLA-based materials are briefly summarized.

SYNTHESIS

PLAs are synthesized by two methods; polycondensation of lactic acids and ring-opening-polymerization (ROP) of lactides, as illustrated in Figure 1 [13].

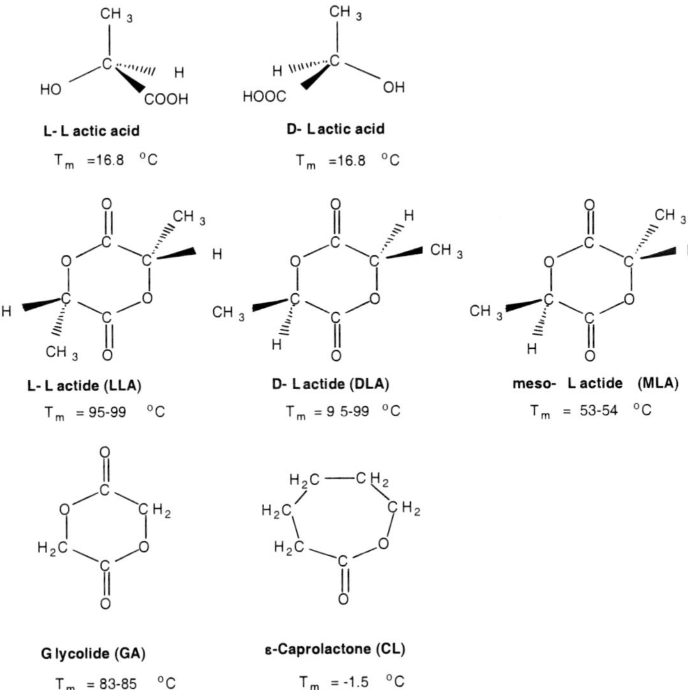

Figure 2. Structure and melting point (Tm) of L- and D-lactic acids, lactides (LAs), and their comonomers [11].

Figure 2 shows the structure of lactic acid, lactide (the cyclic dimer of lactic acid), and comonomers. Here, L-lactide (LLA), D-lactide (DLA), and meso-lactide (meso-LA) are composed of two L-lactic acid (L-lactyl) units, of two D-lactic acid (D-lactyl) units, and an L-lactic acid unit and a D-lactic acid unit, respectively. Racemic lactide or DL-lactide (DLLA) is a 1:1 physical mixture or a 1:1 racemic compound of LLA and DLA, and has a much higher melting temperature (T_m)=124°C compared with those of LLA or DLA (95-99°C).

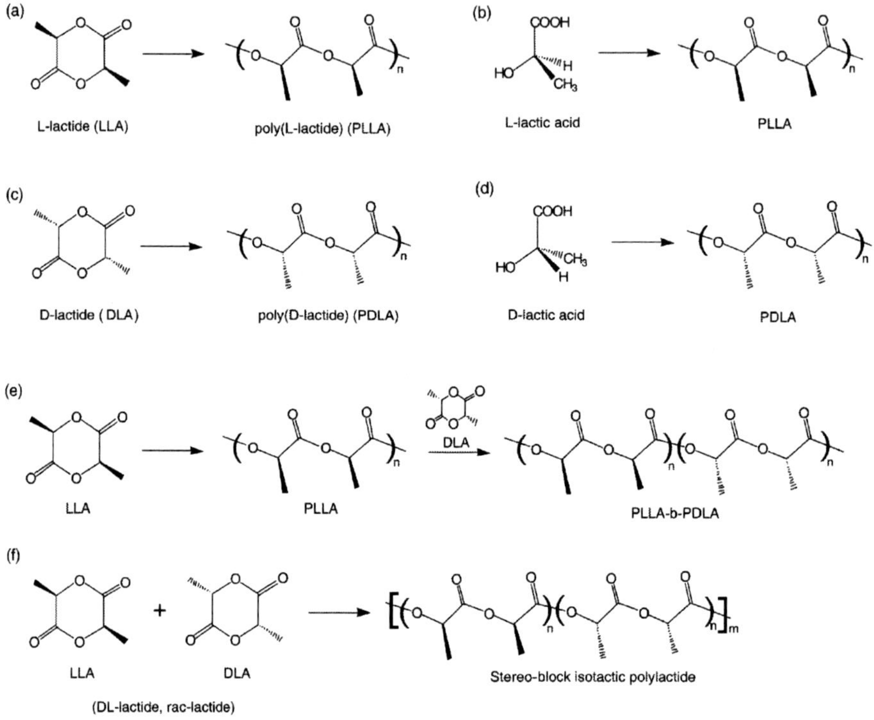

Figure 3. Synthesis and molecular structures of poly(L-lactide), i.e., poly(L-lactic acid) (PLLA) [(a) and (b)], poly(D-lactide), i.e., poly(D-lactic acid) (PDLA) [(c) and (d)], and stereo-block isotactic PLA [(e) and (f)] [13].

Utilizing various initiators, catalysts, and procedures, PLAs having different tactic structures of isomers can be synthesized (Figure 3), whereas copolymerization of lactic acid or lactide with comonomers such as glycolide and ε-caprolactone (Figure 2) will yield a variety of copolymers (Figure 4) from block-type to random-type. Also, ROP of LA lactide with multi-functional coinitiators such as glycerol, pentaerythritol, and poly(vinyl alcohol) gives branched polymers and graft copolymers [11]. These PLAs and copolymers belong to the family of aliphatic polyesters and, therefore, their ester groups are hydrolytically degraded in the presence of water according the following reaction: -COO- + H_2O →-COOH + HO-.

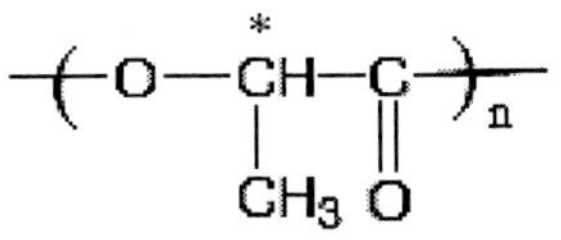

poly(lactide), poly(lactic acid) (PLA)

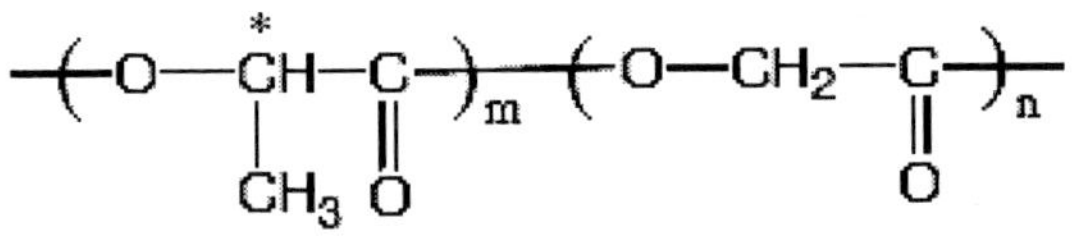

poly(lactide-co-glycolide) [P(LA-GA)]

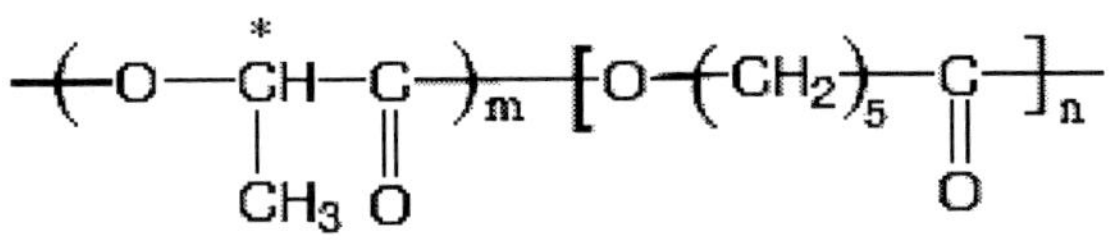

poly(lactide-co-ε-caprolactone) [P(LA-CL)]

Figure 4. Structures of LA homopolymer and linear copolymers.

Chapter 3

APPLICATIONS WITH MANIPULATED DEGRADATION

In industrial and commodity applications, biodegradability is not required because it will reduce the material performance. However, in biomedical, pharmaceutical, and ecological applications, "biodegradability" is functionality. In biomedical and pharmaceutical applications, the degradation rate as well as the mechanical properties of PLA-based materials should be accurately manipulated. For example, the degradation rate of scaffold materials is required to be adjusted to that of tissue regeneration, while the degradation rate of matrices for drug delivery systems should be selected to prolong the effectiveness of drugs. In most of the environmental applications, however, such accurate adjustment of the degradation rate is not required.

Chapter 4

DEGRADATION

Numerous factors affect the degradation mechanism and rate of PLA-based materials. The material and medium factors are summarized in Table 1. When catalytic molecules or substances such as enzymes and alkalis are present in the degradation media or environment, the degradation of PLA-based materials proceeds via a surface erosion mechanism, as schematically illustrated in Figure 5(A) [11,14]. Here, the increase in brightness of the material means a molecular weight decrease. In this surface erosion mechanism, catalytic molecules or ions act only on the surface of materials but will not diffuse into the material so the material is eroded from the surface and the core part of the material remains unchanged. In contrast, in the absence of catalytic molecules or ions as in a phosphate-buffered solution, the degradation of PLA materials takes place via a bulk erosion mechanism [Figure 5(B)]. The hydrolytic degradation mechanism depends on the thickness of biodegradable materials. As shown in Table 2, the critical thickness above which the degradation mechanism changes from bulk erosion to surface erosion depends on the molecular structure of biodegradable (or hydrolyzable) polymers [15]. Also, in the case of PLA, when the thickness of the PLA materials is larger than 2 mm, the hydrolysis-forming oligomers and monomers with a high catalytic effect are entrapped and accumulated in the core part of the materials [16]. This will result in accelerated hydrolytic degradation in the core part (core-accelerated bulk erosion), as shown in Figure 5(C).

**Table 1. Material and medium factors which affect degradation behavior
and rate of PLA-based materials**

Material factors	Medium factors
1. Molecular Structures	Temperature
Molecular weight and distribution	pH
Tacticity (Optical purity) and distribution	Solutes (kinds and concentrations)
Comonomer structure, content, and distribution	Enzymes (kinds and concentrations)
Terminal groups	Microbes (kinds, number, and culture
Branching	conditions)
Crosslinlks	6. Stress or strain
2. Highly-Ordered Structures	
Crystallinity	
Crystalline thickness	
Spherulitic size and morphology	
Orientation	
Hybridization (blends and composites)	
3. Material morphology	
Material shape and dimension	
Porosity and pore size	

**Table 2. Critical thickness (Lcritical) of biodegradable polyesters
where hydrolytic degradation mechanism changes from bulk erosion
to surface erosion [15]**

Polymer	Molecular structure / Example	$L_{critical}$
Poly(anhydride)	$(-R-CO-O-CO-)_n$	75 µm
Poly(ketal)	$(-O-CR^1R^2-O-R^3-)_n$	0.4 mm
Poly(ortho ester)	$[-O-CR^1(OR^2)-O-R^3-]_n$	0.6 mm
Poly(acetal)	$(-O-CHR^1-O-R^2-)_n$	2.4 cm
Poly(α-hydroxycarboxylic acid)	$[-O-CH(CH_3)-CO-]_n$ Poly(lactic acid), Poly(lactide)	7.4 cm
Poly(ε-hydroxycarboxylic acid)	$[-O-(CH_2)_5-CO-]_n$ Poly(ε-caprolactone)	1.3 cm
Poly(amide)	$(-NH-R-CO-)_n$	13.4 m

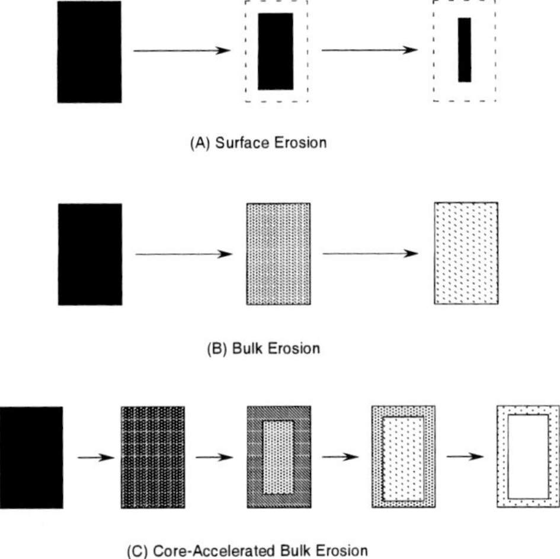

Figure 5. Hydrolytic degradation mechanisms of bulky PLA materials [11,14].

On the other hand, in the case of crystallized PLA materials, numerous spherulites are contained in them. The structure and growth of spherulites are depicted in Figures 6 and 7 [14]. The chains in amorphous regions in crystallized PLA materials are more susceptible to hydrolytic degradation than those in crystalline regions, leaving the chains in crystalline regions intact (Figure 8). The remaining crystalline regions, called "crystalline residues", have the structure of "extended chain crystallites".

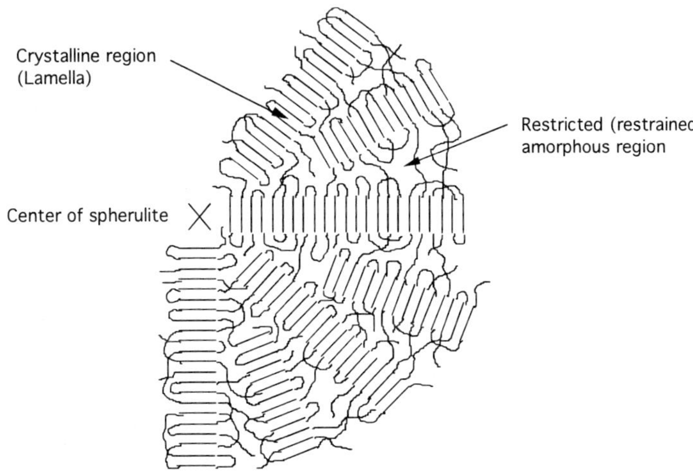

Figure 6. Schematic representation of spherulitic structure [14].

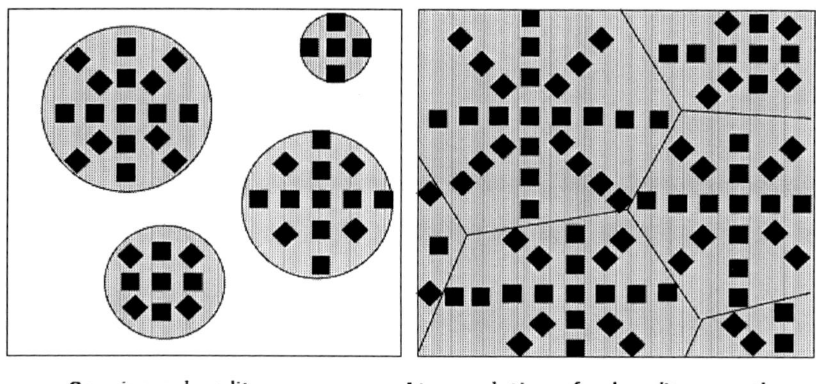

Figure 7. Schematic representation of spherulite growth [14].

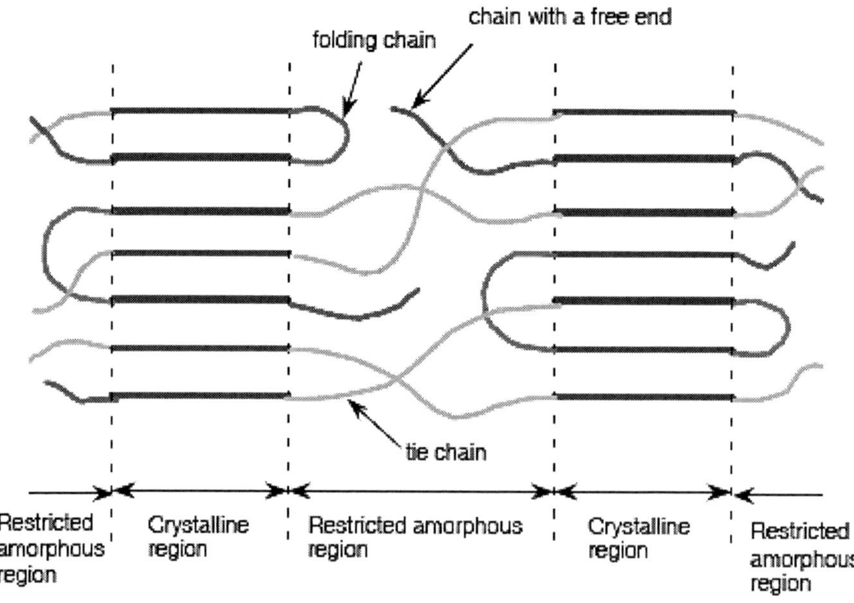

(A) Before hydrolytic degradation.

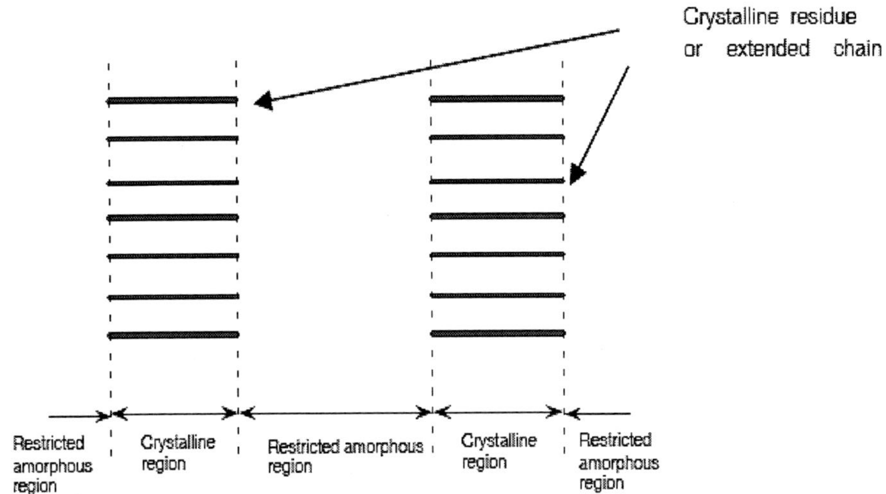

(B) After hydrolytic degradation.

Figure 8. Schematic representation of structures of crystallized PLA material before and after hydrolytic degradation, or of the formation of crystalline residues (or extended chain crystallites), except for proteinase K-catalyzed degradation [14].

Among the factors shown in Table 1, the representative factors which determine the degradation of PLA-based materials are molecular weight, structure and content of comonomer unit, crystallinity, orientation, blending, porosity, pH, temperature, and catalytic molecules or ions. Normally, the incorporation of hydrophilic monomer units or polymers, the presence of catalytic molecules, increasing or decreasing pH, or increasing temperature accelerate the hydrolytic degradation. On the other hand, increasing molecular weight, crystallinity, or degree of orientation reduces the hydrolytic degradation rate. The indexes for hydrolytic degradation or biodegradation are tabulated in Table 3.

Table 3. Indexes for degradation

Material-based indexes	Non-material-based indexes
1. Weight remaining	1. Dissolved organic carbon (DOC) or total organic carbon (TOC)
2. Molecular weight and distribution	
3. Physical properties (e.g., mechanical properties)	2. Biochemical oxygen demand (BOD)
	3. Amount of released carbon dioxide
4. Material morphology	4. Amount of released biogas
	5. pH
	6. Absorbance (Turbidity)

The indexes which can be used for tracing the hydrolytic degradation or biodegradation of materials depend on the erosion mechanism. For example, for a surface erosion mechanism, a significant weight loss is observed at an early stage of degradation, whereas for a bulk erosion mechanism, the weight loss occurs only at a late stage of degradation when a large decrease in molecular weight takes place and, thereby, water-soluble oligomers and monomers are formed. In contrast, molecular weight change is most effective to trace a bulk erosion, but is ineffective in the case of surface erosion. On the other hand, the indexes of BOD, carbon dioxide, and biogas in Table 3 can be used for the biodegradation in the presence of microbes when assimilation of biodegradable materials takes place.

3.1. HYDROLYTIC DEGRADATION

Before discussing enzymatic degradation and biodegradation, the most basic degradation, i.e., hydrolytic degradation without any enzymes or microbes, is treated.

3.1.1. Effects of pH

The presence of alkalis such as NaOH [17-20] or a caffeine base [21] are known to accelerate significantly the hydrolytic degradation of PLA-based materials. In alkaline media, the hydrolytic degradation of PLA-based materials proceeds via the surface erosion mechanism [19], as in enzymatic media. With respect to the effects of molecular structure, we recently investigated the copolymerization effects on alkaline degradation of poly(L-lactide), i.e., poly(L-lactic acid) (PLLA) using PLLA, poly(L-lactide-co-D-lactide) [P(LLA-DLA)] (77:23), poly(L-lactide-co-glycolide) [P(LLA-GA)] (81:19), and poly(L-lactide-co-ε-caprolactone) [P(LLA-CL)] (82:18). It was found that hydrophilicity, which can be traced by water-absorption, is a crucial parameter for determining the alkali-catalyzed hydrolytic degradation rate [22]; the higher the hydrophilicy, the higher the alkaline degradation rate (Figure 9).

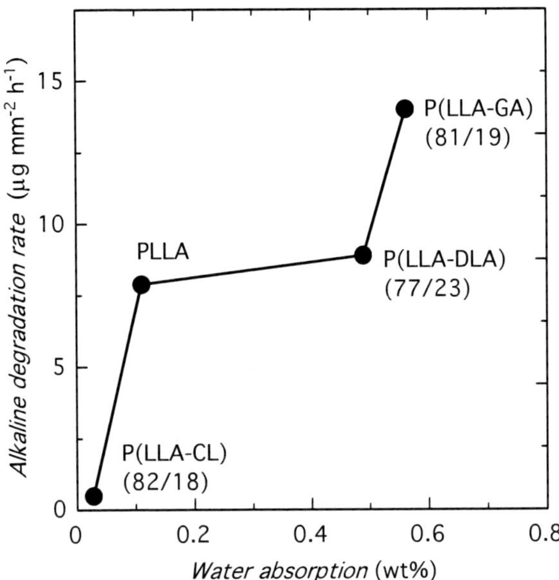

Figure 9. Alkaline hydrolytic degradation rate of amorphous PLLA and copolymers traced by weight loss rate as a function of water absorption [22].

Figure 10 shows the hydrolytic degradation rate of PLLA traced by the weight loss rate as a function of initial crystallinity (X_c) [19], together with that of the proteinase K-catalyzed enzymatic degradation rate [23]. Here, the degradation rate of PLLA at X_c =100% is assumed to be nil. It seems that the alkali-catalyzed hydrolytic degradation rate decreased linearly with X_c, whereas the proteinase K-catalyzed enzymatic degradation rate decreased rapidly from 2.5 to 0.3 μg mm^{-2} h^{-1} with an increase in X_c when it's in the range of 0-30%, and decreased very slowly with an increase in X_c when it exceeds 30%. At X_c from 0 to 30%, most of the amorphous region is composed of a "free" amorphous region, whereas at X_c exceeding 30%, the main species of the amorphous region is a "restricted" amorphous region (Figure 7). Therefore, the results in Figure 10 reveal that in an alkaline solution, the hydrolytic degradation-resistance of "restricted" amorphous regions is similar to that of "free" amorphous region, in marked contrast with the fact that in the presence of proteinase K, a "restricted" amorphous region is much more hydrolytic-resistant compared with a "free" amorphous regions. Cam et al., indicated that X_c is a more dominant parameter to determine the alkaline degradation rate compared with molecular weight [17].

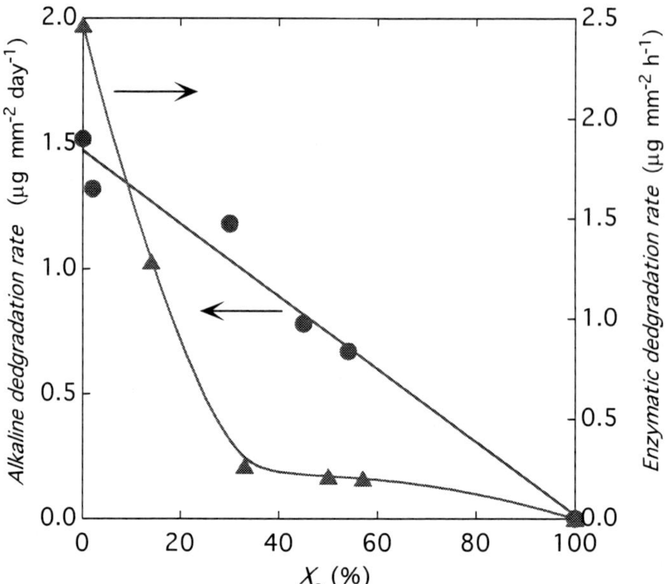

Figure 10. Alkaline hydrolytic degradation rate of PLLA traced by weight loss rate as a function of initial crystallinity (X_c) [19], together with that of proteinase K-catalyzed enzymatic degradation rate [23].

In contrast, acids have no significant catalytic effects on the hydrolytic degradation of high-molecular-weight PLLA even when the pH of the solution was decreased to 1.2 [17] or 2.0 [24]. Tsuji and Nakahara investigated the hydrolytic degradation of PLLA in the presence of two kinds of acids, HCl and DL-lactic acid [24].

The hydrolytic degradation rate constant (k) values of the initially amorphous PLLA film were evaluated from the changes in number-average molecular weight (M_n) according to the following equation:

$$\ln M_n(t_2) = \ln M_n(t_1) - k\,(t_2 - t_1) \tag{1}$$

where $M_n(t_2)$ and $M_n(t_1)$ are the M_n values at the hydrolytic degradation times of t_2 and t_1, respectively. The estimated k values were 3.0×10^{-3} and 2.4×10^{-3} day^{-1} at pH 2.0 in the HCl and DL-lactic acid solutions, respectively. These k values are very similar to 2.6×10^{-3} day^{-1} at pH 7.4 in a phosphate-buffered solution. The similar k values and the negligibly small weight loss after acidic hydrolytic degradation for 300 days reflect the fact that the hydronium ions and the lactic acid oligomers and monomers have no significant catalytic effect on the hydrolytic degradation of the PLLA films. An increment in the initial X_c of PLLA film increased the hydrolytic degradation rate in the HCl solution, consistent with that in the phosphate-buffered solution, whereas increasing the initial crystallinity of PLLA film did not alter the hydrolytic degradation rate in the DL-lactic acid solution.

In the case of PLLA crystalline residues (Figure 8), the deviation of pH from 7 to both higher and lower values accelerated the hydrolytic degradation [25]. Figure 11 shows the hydrolytic degradation rate constant (k) values estimated from M_n as a function of pH [25]. For the estimation of the k values, the following equation was used:

$$Mn(t2)=Mn(t1)-k (t2 - t1) \qquad (2)$$

It is seen that both the hydronium ion and the hydroxide ion accelerate the hydrolytic degradation of PLLA crystalline residues. Other plots of k as a function of hydronium ion concentration [H$^+$] or hydroxide ion concentration [OH$^-$] (not shown here) indicated that hydroxide ion has a higher accelerating effect on hydrolytic degradation.

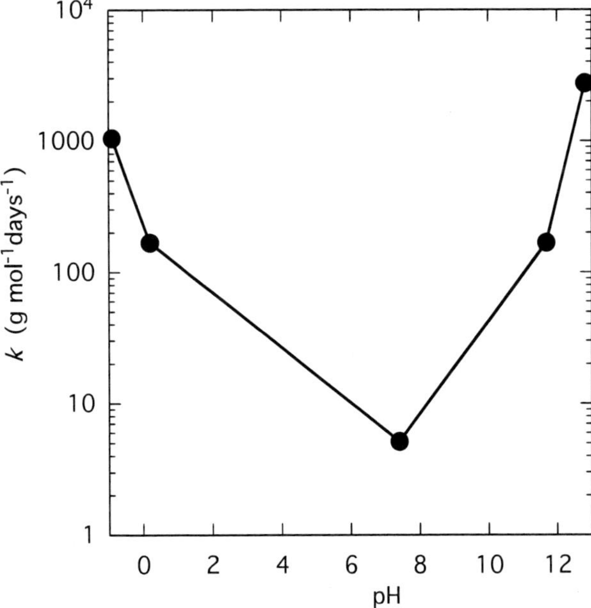

Figure 11. The hydrolytic degradation rate (k) of PLLA crystalline residues (or extended chain crystallites) as a function of pH [25].

3.1.2. Effects of Temperature

In sections 3.1.2-3.1.7, various effects are stated for the hydrolytic degradation of PLA only in neutral media, as in a phosphate-buffered solution, to exclude the effects of pH. The hydrolytic degradation temperature (T_{HD}) of PLLA and crystallizable PLA can be divided into three ranges; (1) T_{HD} < glass transition temperature (T_g), (2) $T_g \leq T_{HD} < T_m$, and (3) $T_m \leq T_{HD}$. Here, the T_g and T_m of high molecular weight and optically pure PLLA are in the ranges of 55-60°C and 175-180°C, respectively. The hydrolytic degradation in the temperature ranges of (1) and (2) proceed in a solid state, whereas that in the temperature range of (3) takes place in a molten state. In the former case, at an early stage, hydrolytic degradation is restricted to occurring in amorphous regions. That is, hydrolytic degradation takes place at a higher rate in amorphous regions and at a very slow rate in crystalline regions. As a result, crystalline residues are formed at a late stage of degradation. For the temperature ranges of (1) and (2), it is reported that the hydrolytic degradation rate of poly(L-lactide-co-glycolide), estimated from the tensile strength, increased dramatically above T_g (36°C), reflecting that segmental mobility affects the degradation rate [26]. In contrast, in the latter case, all the PLLA chains are molten or in an amorphous state and, therefore, degradation proceeds homogeneously in the materials [27]. The hydrolytic degradation rate constants (k) of PLLA [27], poly[(R)-3-hydroxybutyrate] [R-P(3HB)] [28], and PCL [29] were estimated from the M_n change using equation (1), and their Arrhenius plots are shown in Figure 12.

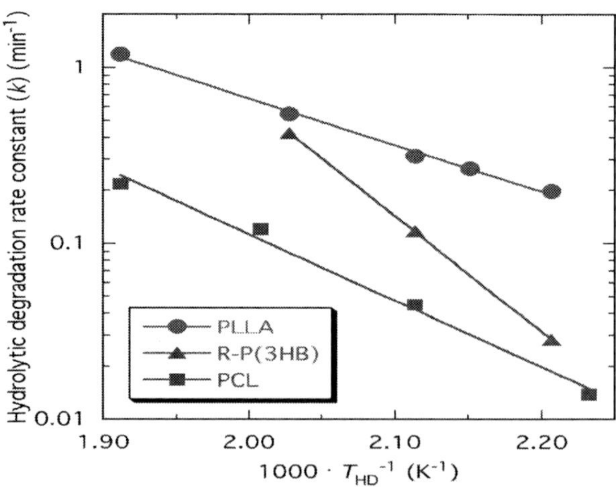

Figure 12. Arrhenius plot of PLLA, R-P(3HB), and PCL in the melt [27-29].

PLLA has the highest k values in the temperature range of 180-220°C among these aliphatic polyesters. From Figure 12, the dependence of k on T_{HD} was obtained as follows:

$$k \ (\text{min}^{-1}) = 1.39 \times 10^5 \exp(-6.12 \times 10^3 / T_{HD}) \ (\text{PLLA}) \tag{3}$$

$$k \ (\text{min}^{-1}) = 3.95 \times 10^6 \exp(-8.69 \times 10^3 / T_{HD}) \ (\text{PCL}) \tag{4}$$

$$k \ (\text{min}^{-1}) = 8.86 \times 10^{12} \exp(-1.51 \times 10^4 / T_{HD}) \ [\text{R-P(3HB)}] \tag{5}$$

The activation energy values (ΔE_h) for the hydrolytic degradation in the melt were obtained from equations (3)-(5). The ΔE_h value for PLLA of 12.2 kcal mol^{-1} (50.9 kJ mol^{-1}) in the temperature range of 180-250 °C is lower than 17.3 kcal mol^{-1} (72.2 kJ mol^{-1}) for PCL in the temperature range of 175-250°C, and 30.0 kcal mol^{-1} (126 kJ mol^{-1}) for R-P(3HB)) in the temperature range of 180-220°C [27-29], and much lower than 55.7 kcal mol^{-1} for an aromatic polyester, PET, in the temperature range of 250-280°C [30].

In the temperature ranges of (1) and (2), at the late stage of hydrolytic degradation, the chains in amorphous regions are completely removed and only the crystalline residues (or extended chain crystallites) remain. The crystalline residues contain a negligibly small amount of amorphous chains at each chain terminal. In such a case, k is not influenced by the condition of whether T_{HD} is higher or lower than T_g. We prepared PLLA crystalline residues by accelerated hydrolytic degradation of crystallized PLLA films at 97°C for 40 hours [31,32]. Figure 13 shows the M_n change of PLLA crystalline residues during the course of hydrolytic degradation at pH 7.4 and different temperatures [33]. The M_n of PLLA crystalline residues decreased linearly with degradation time according to equation (2), irrespective of temperature. This result is in agreement with the result at 37°C and pH −0.9, 0.2, 11.7, and 12.8, but in marked contrast with the hydrolytic degradation of PLLA materials at an early stage, wherein the M_n decreased exponentially with degradation time according to equation (1). This finding shows that the degradation of PLLA crystalline residues proceeds only at the chain terminals on their surface. Also, the M_n decrease rate was elevated by increasing the degradation temperature.

The Arrhenius plots of k values obtained at different degradation temperatures in Figure 13 yields the ΔE_h values of PLLA crystalline residues [33].

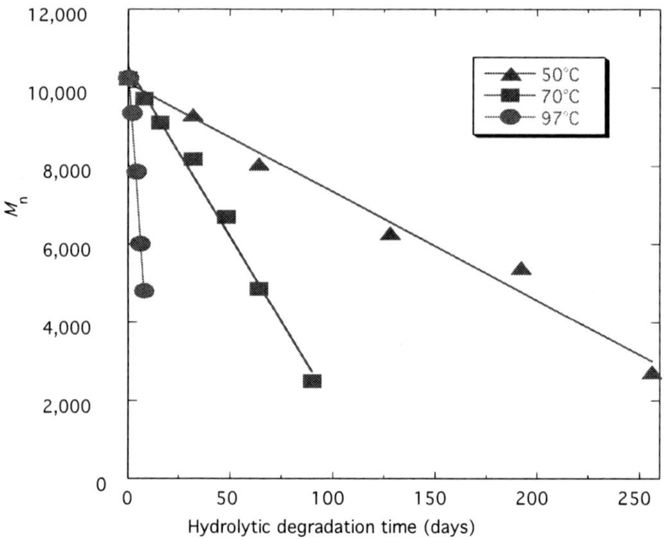

Figure 13. The M_n change of PLLA crystalline residues during the course of hydrolytic degradation at pH 7.4 and different temperatures [33].

The obtained value of 18.0 kcal mol^{-1} (75.2 kJ mol^{-1}) in the temperature range of 50-97°C is higher than 12.2 kcal mol^{-1} (50.9 kJ mol^{-1}) for PLLA hydrolytic degradation in the melt [27], revealing that the PLLA chains in the crystalline residues are much more hydrolytic degradation-resistant than those in an amorphous state or in the melt. However, the ΔE_h value of PLLA crystalline residues is slightly lower than 20.0 and 19.9 kcal mol^{-1} (83.7 and 83.3 kJ mol^{-1}, respectively) reported respectively for PLLA and poly(DL-lactide), i.e., poly(DL-lactic acid) (PDLLA) microspheres as solids in the temperature range below T_g (21-45°C) [17].

3.1.3. Effects of Molecular Structures

Among the effects of molecular structure, those of the monomer unit structure and the molecular weight of a polymer are crucial. In sections 3.1.3-3.1.6, the hydrolytic degradation media have neutral pH and the temperature is 37°C, unless otherwise specified. The incorporation of comonomer units such as glycolide (GA) and ε-caprolactone (CL) accelerates the hydrolytic degradation of PLA-based copolymers, irrespective of the hydrophilicity of the comonomer. This is in marked contrast to alkaline degradation, wherein the hydrophilicity of the comonomer has a dominant effect on hydrolytic degradation [22]. Figure 14 shows the M_n change of amorphous PLLA, P(LLA-DLA)(77:23), P(LLA-GA)(81:19), and P(LLA-CL)(82:18) with respect to hydrolytic degradation in a phosphate-buffered solution [34].

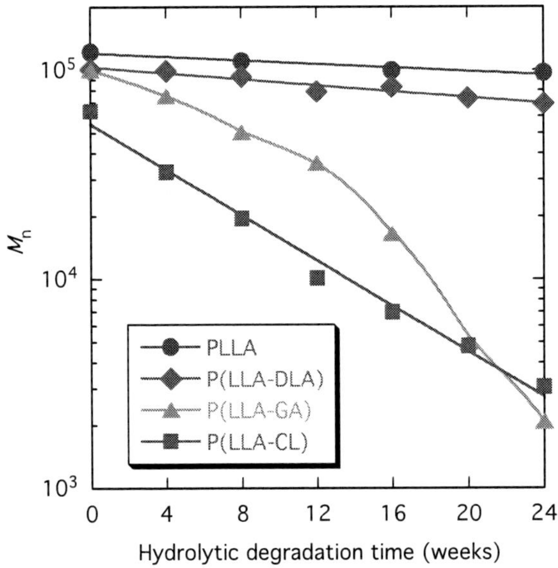

Figure 14. M_n change of amorphous PLLA, poly(L-lactide-co-D-lactide) [P(LLA-DLA)] (77:23), poly(L-lactide-co-glycolide) [P(LLA-GA)] (81:19), and poly(L-lactide-co-ε-caprolactone) [P(LLA-CL)] (82:18) with respect to hydrolytic degradation in phosphate-buffered solution [34].

The k values were estimated from the M_n change in Figure 14 using equation (1). For the estimation of the k value of P(LLA-GA) film, the M_n data for the period of 0-16 weeks (not for the period of 0-24 weeks) were used, because the slope of log M_n changed at 16 weeks. The estimated k values for PLLA, P(LLA-DLA), P(LLA-GA) films were 1.11×10^{-3}, 2.39×10^{-3}, 1.61×10^{-2}, and 1.81×10^{-2} day^{-1}. The k values of PLLA and P(LLA-DLA) films are comparable to 2.59×10^{-3} day^{-1} (0-52 weeks) reported for an amorphous-made PLLA film ($M_w = 1.2 \times 10^6$ g mol^{-1}, $M_w/M_n = 2.6$) [35], and that of P(LLA-GA) film is very similar to 1.91×10^{-2} day^{-1} reported for a P(LLA-GA)(75:25) scaffold ($M_n = 1.81 \times 10^5$ g mol^{-1}, $M_w/M_n = 1.79$) [36]. The estimated k value for P(LLA-CL) film is comparable to 3.16×10^{-2} day^{-1} reported for P(LLA-CL)(60:40) ($M_n = 7.4 \times 10^4$ g mol^{-1}, $M_w/M_n = 2.0$)

[37]. The rather higher k value reported for P(LLA-CL)(60:40) may be due to the higher fraction of CL units and thickness of the copolymer specimen compared with those in the present study.

The k values were higher for P(LLA-GA) and P(LLA-CL) films than those for PLLA and P(LLA-DLA) films. The higher hydrolytic degradation rates of P(LLA-GA) and P(LLA-CL) films are partly attributable to their higher chain mobility. This is evidenced by the fact that the T_g values of P(LLA-GA) and P(LLA-CL) films (56 and 27°C, respectively) were lower than those of PLLA and PDLLA films (60 and 58°C, respectively) [34]. Such higher chain mobility of P(LLA-GA) and (PLLA-CL) films facilitates faster water supply, and thereby enhances hydrolytic degradation. Furthermore, in the case of P(LLA-GA), the higher hydrophilicity and lower steric hindrance of GA units compared with those of the L-lactide (LLA) unit should have enhanced the water diffusion and concentration of P(LLA-GA) film, resulting in rapid hydrolytic degradation. Incorporation of D-lactide (DLA) in PLLA chains is known to increase the hydrolytic degradation rate [38,39]. We have recently found that even the incorporation of a very small amount of DLA units (1.2%) in PLLA chains significantly elevates the hydrolytic degradation rate [40].

Although molecular weight has a very small effect on the hydrolytic degradation rate of PLA-based materials with molecular weights exceeding 8×10^4 g mol^{-1} [40], the effect becomes significant for molecular weights below 8×10^4 g mol^{-1} [41]. The latter result can be explained by four factors: (1) the elevated molecular mobility, (2) the increased density (number per unit mass) of hydrophilic terminal carboxyl and hydroxyl groups, (3) the increased density of catalytic terminal carboxyl groups, (4) the higher probability of the formation of water-soluble oligomers and monomers upon hydrolytic degradation. Factors (1) and (2) increase the water diffusion rate and content, enhancing the hydrolytic degradation. Factors (1)-(4) should be dramatic for PLA at a molecular weight lower than 1×10^4 g mol^{-1}.

3.1.4. Effects of Highly Ordered Structures

Figure 15(C) shows the molecular weight distribution change of crystallized PLLA hydrolytically degraded in a phosphate-buffered solution. As depicted in Figure 8, the chains in the amorphous regions are selectively degraded and removed, and thereby the crystalline residues (extended chain crystallites) are formed. The peak at around 1×10^4 g mol^{-1} is ascribed to one fold of the thickness of the PLLA crystalline residues. Among the effects of highly ordered structures, that of crystallinity is reported to have a crucial influence on the hydrolytic degradation of PLLA in neutral media [11].

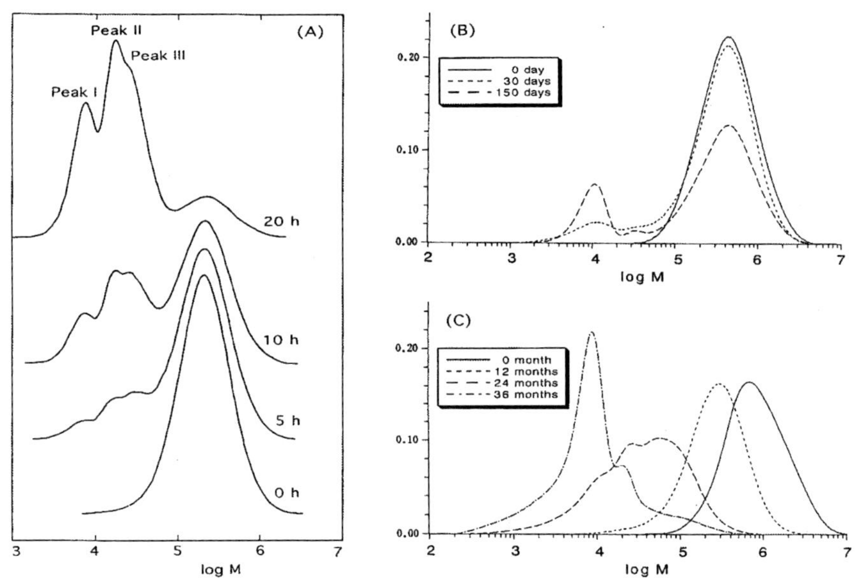

Figure 15. Molecular weight distribution curves of crystallized PLLA film hydrolytically degraded in (A) enzymatic, (B) alkaline, and (C) phosphate-buffered solutions [11].

Figure 16 shows the M_n change of PLLA films having different crystallinity (X_c) values in a phosphate-buffered solution [35]. Surprisingly, the crystallization of PLLA or the elevated crystallinity of PLLA enhances its hydrolytic degradation [35,42-44]. This is in marked contrast with the results for proteinase K-catalyzed enzymatic degradation [23] and alkaline degradation [18,19], wherein the degradation rate is higher for PLLA materials having a lower crystallinity. The result here can be explained as follows. Upon the crystallization of PLLA, the hydrophilic terminal groups (-OH and –COOH) and the catalytic terminal group (-COOH) are condensed in the amorphous region between the crystalline regions, as depicted in Figure 17 [14,35]. The open and closed circles in the figure refer to hydroxyl and carboxyl groups, respectively, or vice versa. The high densities (numbers per unit mass) of terminal groups will cause loose chain packing in the amorphous region between the crystalline regions compared with those in the completely amorphous film. Such loose chain packing and high densities of the hydrophilic terminal groups enhance the diffusion of water molecules and increase the water content.

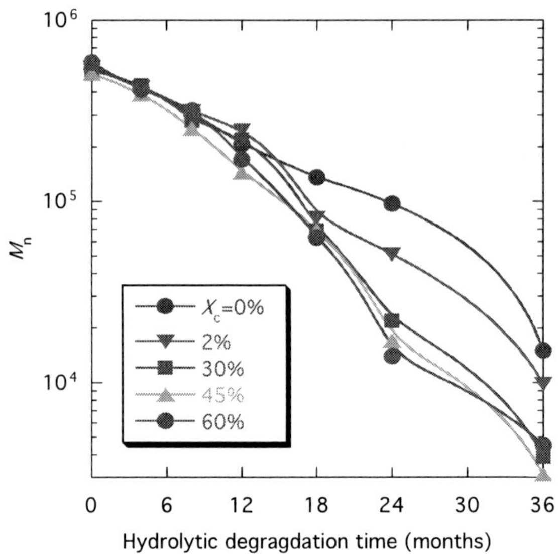

Figure 16. M_n change of PLLA films having different crystallinity (X_c) values in phosphate-buffered solution [35].

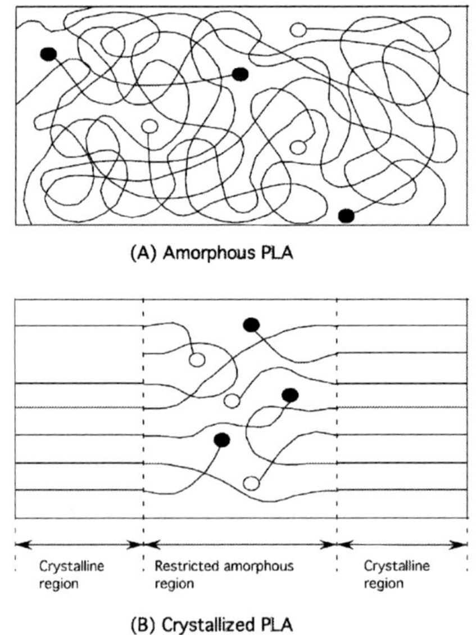

Figure 17. Schematic representation of completely amorphous and crystallized PLLA [14,35].

The elevated water supply rate and content and the catalytic effect increased by the high density of the carboxylic group will synergize to remarkably accelerate hydrolytic degradation of crystallized PLLA films. In contrast to the enzymatic degradation of PLLA [45,46], orientation has a relatively small effect on hydrolytic degradation in a neutral medium [47].

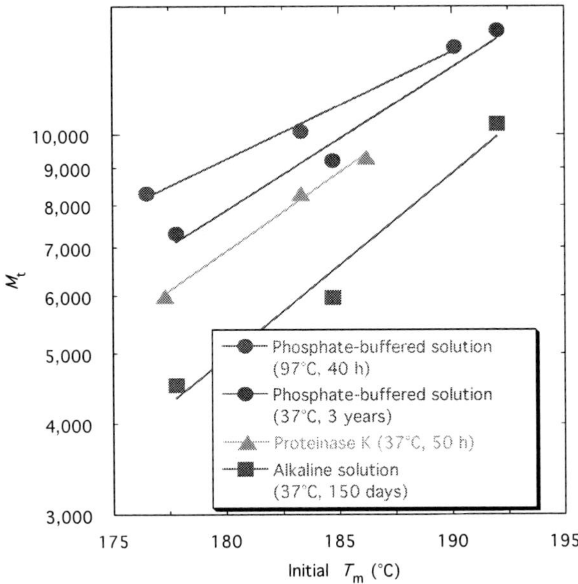

Figure 18. Initial T_m before hydrolytic degradation and peak molecular weight (M_t) after hydrolytic degradation [32].

The initial crystalline thickness affects the hydrolytic degradation behavior of PLLA materials especially at a late stage. Figure 18 shows the initial T_m before hydrolytic degradation and peak molecular weight (M_t) after hydrolytic degradation [32]. This shows that the molecular weight of crystalline residues depends on the initial crystalline thickness, while the molecular weight of crystalline residues formed by hydrolytic degradation depends on the media and other conditions.

The molecular weight of Peak I (the lowest) ($M_{n,s}$) can be converted to the lamella thickness (L_c) of PLLA after hydrolytic degradation using the following equation assuming the PLLA chains take 10_3 helices in the α-form unit cell having a dimension of c=2.78 (or 2.88) nm (fiber axis) [48,49]:

$$L_c \text{ (nm)} - 0.278 \text{ (or } 0.288) \times M_{n,s} / 72.1 \tag{6}$$

where 72.1 is the mass per mole of the lactyl unit (half of the lactide unit). The following equations show the relationship between final T_m and L_c:

$$T_m \text{ (K)} = 471 \, [1\text{-}1.59/L_c \text{ (nm)}] \quad \text{(Degradation in phosphate-buffered solution)} \quad (7)$$

$$T_m \text{ (K)} = 472 \, [1\text{-}1.46/L_c \text{ (nm)}] \quad \text{(Degradation in the presence of proteinase K)} \quad (8)$$

Here, the c values of 2.78 nm and 2.88 nm, respectively, were used to calculate the L_c values in equations (7) and (8). The Thomson-Gibbs expression between T_m and L_c is given by the following equation [50]:

$$T_m = T_m^0 \, (1\text{-}2\sigma/\Delta h^0 \rho_c L_c) \quad (9)$$

where σ, Δh^0, and ρ_c are the specific fold surface free energy, heat of fusion (per unit mass), and crystal density, respectively. The comparison between (7) or (8) and (9) gives the T_m^0 values of 198 and 199°C as the T_m^0 values, which are slightly lower but almost in agreement with those obtained from the Hoffman-Weeks procedure for the melt-crystallized PLLA specimens by Kalb and Pennings (215°C) [51] and by ourselves (212°C) [52]. This agreement on T_m^0 estimated by different methods also confirms our assumption that $M_{n,s}$ is the molecular weight of one fold of the PLLA chain in the crystalline regions.

3.1.5. Effects of Blending and Additives

Five or more factors for a second polymer or an additive are anticipated to influence the hydrolytic degradation rate and behavior of PLA-based materials in a neutral media; (1) miscibility and dispersibility, (2) hydrophilicity, (3) acidity or basicity, (4) molecular weight, (5) size and shape of the polymer domain (if the second polymer is immiscible with the first polymer) or additive. For example, the addition of hydrophilic polymers such as poly(vinyl alcohol) (PVA) [53,54] and poly(ethylene glycol) (PEG) [55], basic additives such as thioridazine [56], or catalytic additives such as lauric acid [57] and lactide [42], is known to enhance the hydrolytic degradation of PLLA in a neutral medium. Normally, the hydrolytic degradation rate constant (k), estimated using equation (1), is in the range of 2.2-3.4x10^{-3} day^{-1} (0-12 months) [35,58]. The addition of PVA and lauric acid respectively increased the k value to 9.4x10^{-3} day^{-1} [54] and 0.03 day^{-1} [56]. A peculiar example is the addition of poly(D-lactide) [i.e., poly(D-lactic acid) (PDLA)] to PLLA. In the blends, the interaction between a PLLA chain and a

PDLA chain is much higher than that between PLLA chains or PDLA chains, resulting in a stereocomplex formation [59]. Such strong interaction disturbs the diffusion of water into the material and lowers the hydrolytic degradation rate. For example, as-cast pure PLLA and PDLA films had k values of 1.41×10^{-3} and 2.14×10^{-3} day^{-1}, respectively, whereas the k value of their 1:1 blend film was 7.3×10^{-4} day^{-1} [59]. That is, the 1:1 blend film has a k value one order lower than those of nonblended films.

3.1.6. Effects of Material Shape

Both thinning materials and pore formation decrease the average length required for the hydrolysis-forming water-soluble oligomers and monomers to diffuse outside and for water molecules to diffuse inside the material. Therefore, such shape changes will increase the transfer rates of lactic acid oligomers and monomers and water molecules. The increased transfer rate of lactic acid oligomers and monomers will decrease their catalytic effects for hydrolytic degradation, whereas the elevated transfer rate of water will elevate the hydrolytic degradation rate. The reported result that porous PLLA had a lower hydrolytic degradation rate than that of nonporous PLLA [60] indicated that the former effect prevailed over the latter one.

3.1.7. Applications for Recycling of PLA

The hydrolytic degradation process can be used for the recycling of PLLA to its monomer, L-lactic acid. In this process, a high yield in a short period of time is crucial. By elevating the degradation temperature, a high degradation rate of PLLA may be attained. However, too high a temperature such as 300°C causes racemization and degradation of the formed L-lactic acid [27]. By decreasing the degradation temperature, the issues of racemization and degradation of L-lactic acid can be avoided. However, a hydrolytic temperature below the T_m of PLLA will cause the formation of crystalline residues (or extended chain crystallites). As stated above, the hydrolytic degradation rate of the crystalline residues is much lower compared with that of amorphous chains and, therefore, requires a long period of time to become an L-lactic acid, e.g., 14 days at 97°C [33]. The most appropriate temperature range for PLLA recycling was reported to be 180-250°C, where no significant racemization or degradation of formed L-lactic acid takes place [27,61].

**Table 4. Commercially available enzymes which catalyze
hydrolytic degradation of PLA-based materials [64, 96-98]**

PLA	Origin	Manufacturer	Type / Name (Trade name)	Researcher [Reference]
PLLA	Animal pancreas	Amano Pharm. Co. (Nagoya, Japan)	Protease / Pancreatin F	Oda et al. [98]
	Aspergillus oryzae	Amano Pharm. Co. (Nagoya, Japan)	Protease / Protease A "Amano" (Activity is low)	Oda et al. [98]
	Bacillus lentus	Novo Nordisk Bioind. Co. (Chiba, Japan)	Protease / Savinase 16.0L Protease / Esperase 8.0L	Oda et al. [98]
		Kyowa Enzymes Co. (Tokyo, Japan)	Protease / Purafect 4000L	
	Bacillus licheniformis	Kyowa Enzymes Co. (Tokyo, Japan)	Protease / Alcalase 2.5L DX Protease / Alkaline protease GL-440	Oda et al. [98]
	Bacillus subtilis	Hankyu Bioind. Co. (Osaka, Japan)	Protease / Orientase Y Protease / Orientase 5BL Protease / Orientase 22BF HB	Oda et al. [98]
		Daiwa Kasei Co. (Osaka, Japan)	Protease / Protin A	
		Amano Pharm. Co. (Nagoya, Japan)	Protease / Proleather	
	Pineapple cannery	Amano Pharm. Co. (Nagoya, Japan)	Protease / Bromelain F (Activity is very low)	Oda et al. [98]
	Tritirachium album	E.g., Sigma-Aldrich Co. (MO, USA)	Protease / Proteinase K /	Williams [64]
	Unknown	Nagase Biochem. Ind. Co. (Fukuchiyama, Japan)	Protease / Bioprase AL-15FG	Oda et al. [98]
PDLLA	*Fusarium solani*	CORVAS International N.V. (Gent, Belgium)	Esterase / Cutinase	Ivanova et al. [97]
	Rhizopus delemer	Sigma-Aldrich Co. (MO, USA)	Lipase	Fukuzaki et al. [96]
	Hog pancreas	Sigma-Aldrich Co. (MO, USA)	Lipase	Fukuzaki et al. [96]
	Porcine liver	Sigma-Aldrich Co. (MO, USA)	Esterase	Fukuzaki et al. [96]
	Wheat germ	Sigma-Aldrich Co. (MO, USA)	Lipase	Fukuzaki et al. [96]

3.2. ENZYMATIC DEGRADATION

Enzymatic and microbial degradation of PLA has been well documented in review articles by Tokiwa and coworkers [62,63]. Some enzymes, proteases and lipases, have been reported to catalyze the chain cleavage of PLA. In 1981, Williams reported that proteinase K, one of the proteinases, can catalyze the hydrolytic degradation of PLA [64]. Later, Gross and McCarthy et al. [45,65-69], Nagata et al. [70], Li et al. [71-74], Shirahama and Yasuda et al. [75-80], and Tsuji et al. [20,22,23,25,46,54,81-94] investigated the various parameter effects of molecular structures, highly-ordered structures, and additives. Recently, Yamashita et al. investigated the enzymatic degradation of PLLA using a quartz crystal microbalance. It was found that the degradation rate is determined by the amount of adsorbed proteinase K, and that the adsorption of proteinase K is irreversible despite a lack of a binding domain [95]. The commercially available enzymes, which can catalyze the hydrolytic degradation of PLA, are summarized in Table 4 [64,96-98]. It should be noted that in ref. [96] relatively low molecular weight PLA (M_n of about 2,200) was used to test enzymatic degradability and that most lipases therein have no significant catalytic effect on hydrolytic degradation for high-molecular-weight PLA.

3.2.1. Effects of Molecular Structures

As proteinase K and other proteinases catalyze the hydrolytic degradation of L-type poly(α-amino acid), in the case of lactide or lactic acid homopolymers, the L-lactyl unit sequence length is the most crucial factor for determining enzymatic degradability; the longer the L-lactyl unit sequence length, the higher the enzymatic degradation rate [22,65,67,71,73,74,75-80,83]. Reeve et al. reported that PLLA has the highest proteinase K-catalyzed enzymatic degradation rate, and that the enzymatic degradation rate decreased with an increase in D-lactyl unit content [65]. Tsuji and Miyauchi found that at least four L-lactyl unit sequences are required for proteinase K-catalyzed hydrolytic degradation of PLA and that the enzymatic degradation rate decreases with an increase in the molecular weight of PLLA [83]. The latter finding suggests that proteinase K catalyzes the cleavage of L-lactyl unit sequences at the chain terminals. Tsuji and Tezuka investigated the effects of the copolymerization of GA, DLA, and CL units on proteinase K-catalyzed enzymatic degradation of L-lactide-based polymers, and indicated that, similar to alkaline degradability, enzymatic degradability becomes lower with the incorporation of hydrophobic comonomers such as CL (Figure 19) [22].

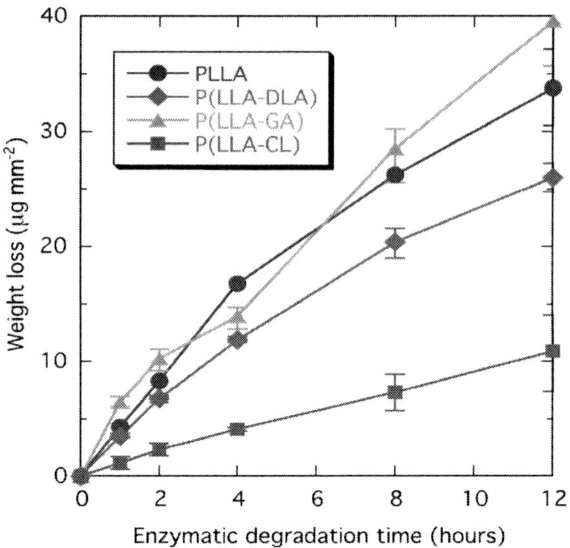

Figure 19. Weight loss of amorphous PLLA, poly(L-lactide-co-D-lactide) [P(LLA-DLA)] (77:23), poly(L-lactide-co-glycolide) [P(LLA-GA)] (81:19), and poly(L-lactide-co-ε-caprolactone) [P(LLA-CL)] (82:18) with respect to proteinase K-catalyzed enzymatic degradation [22].

In our study [22,54,83], completely amorphous specimens were used for the experiments to exclude the effects of highly ordered structures such as crystallinity, and to thereby investigate the pure effects of the molecular structures.

3.2.2. Effects of Highly-Ordered Structures

As expected, with an increase in crystallinity, the proteinase K-catalysed enzymatic degradation rate decreases [23,67,71,82]. This is due to the fact that enzyme molecules cannot diffuse into the rigid crystalline regions. An interesting finding for the proteinase K-catalyzed enzymatic degradation of crystallized PLLA specimens is that the enzyme has a selective activity toward tie chains and chains with a long free end in amorphous regions [Figure 15(A)] [11]. In other words, folding chains in amorphous regions are enzymatic degradation-resistant [25,82]. This is evidenced by the molecular weight change during enzymatic degradation. Here, the peak ascribed to the undegraded core part remains at the same position, whereas specific low-molecular-weight peaks appear at the molecular weights of 1, 2, and 3×10^4 g mol^{-1}, which are attributed to one-, two-, and three-fold crystalline thickness [Figure 15(A)]. The fact that peaks with the molecular weights of two- and three-fold crystalline thickness remains for as long as 70 hours, even when the height of the initial main peak decreases to 20% of its initial value, strongly suggests that the folding chains in the restricted amorphous region between the crystalline regions are enzymatic degradation-resistant in the presence of proteinase K compared with tie chains and chains with free ends. The lowest molecular weight of a specific peak ascribed to the PLLA crystalline thickness after enzymatic degradation increases with the initial crystalline thickness before enzymatic degradation or the crystallization temperature for material preparation [82]. With respect to proteinase K-catalyzed enzymatic degradation of PLLA single crystals, Iwata and Doi concluded that single crystals were degraded at the disordered chain-packing region of crystal edges rather than the chain-folding surfaces of single crystals [99]. This is consistent with our result that folding chains are enzymatic degradation-resistant.

Another interesting finding is that the cleavage of chains with a free end is not catalyzed by proteinase K when the chain length from the end of the crystalline region to the free end becomes very short [25]. We prepared crystalline residues (or extended chain crystallites) from crystallized PLLA specimens by accelerated non-enzymatic hydrolytic degradation in a phosphate-buffered solution at 97°C [25,31,32] and carried out the proteinase K-catalyzed enzymatic degradation of the crystalline residues. No change in molecular weight distribution was observed even when the enzymatic degradation was continued for 50 days.

As stated earlier, the enzymatic degradability of folding chains and the very short chains with a free end is very low. We call the amorphous region sandwiched by two crystalline regions in the spherutlites the "restricted" amorphous region (or "restrained" amorphous region) (Figures 6, 7, and 17) [14,35] and the amorphous region outside the spherulites the "free" amorphous region. Figure 20 shows the proteinase K-catalyzed enzymatic degradation rate of PLLA specimens crystallized at different temperatures for a long period which have no crystallizable "free" amorphous region [82]. That is, in these specimens amorphous regions are "restricted" amorphous regions. The extrapolation of the data for different X_c values to zero X_c gives the enzymatic degradation rate of "restricted" amorphous regions (Figure 20) [82]. The estimated value of 0.36 µg mm^{-2} h^{-1} is much lower than the enzymatic degradation rate of 2.47 µg mm^{-2} h^{-1} of "free" amorphous region [23,82]. This result demonstrates that a restricted amorphous region is much more enzymatic degradation-resistant compared to a free amorphous region. However, no such difference in degradability was observed for the proteinase K-catalyzed enzymatic degradation of P(LLA-

DLA)(95:5) [89]. This large hydrolyzability difference of "free" and "restricted" amorphous regions between PLLA films and P(LLA-DLA)(95:5) films can be explained by the large difference in the chain packing or mobility between "free" and "restricted" amorphous regions, as evidenced by their respective T_g values, 61 and 54°C for PLLA and P(LLA-DLA) (95:5) films crystallized at 100°C. Also, Kikkawa et al. found that crystallized PLLA thin films have two amorphous regions, identified as free and restricted amorphous regions [100].

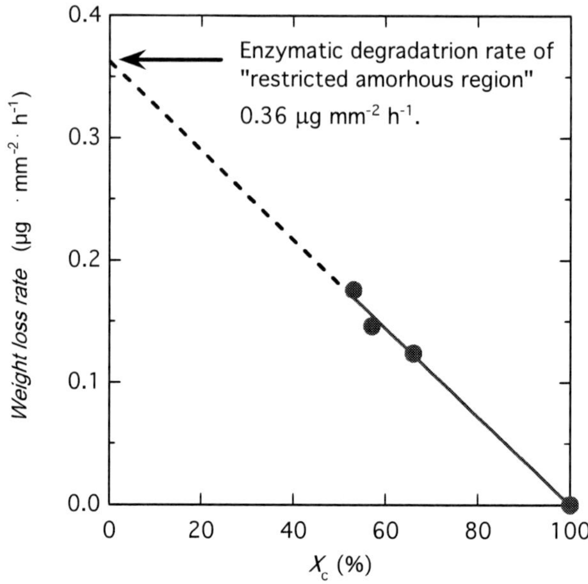

Figure 20. Proteinase K-catalyzed enzymatic degradation rate of PLLA films crystallized at different temperatures (which have no free amorphous region) as a function of X_c [82].

On the other hand, both uniaxial and biaxial orientation of PLLA materials lowers the proteinase K-catalyzed enzymatic degradation rate [45,46,92]. This result is attributable to the fact that proteinase K cannot attach to extended or strained PLLA chains, or cannot catalyze the cleavage of such chains. Figure 21 shows the molecular weight distribution changes of biaxially oriented and de-oriented PLLA films (abbreviated as PLLA-O and PLLA-C, respectively) during proteinase K-catalyzed enzymatic degradation [46]. For the de-oriented PLLA-C films, the formation of specific low-molecular-weight multiple peaks in the range of 2×10^3-3×10^4 g mol^{-1} reflect the fact that the crystalline residues were accumulated during enzymatic degradation, in marked contrast with the results for alkaline degradation. The multiple low-molecular-weight specific peaks correspond to one, two, and three folds of PLLA chain in the crystalline regions. These specific peaks were formed as a result of a much higher degradation-resistance of the folding chains in the presence of proteinase K than that of the tie chains and the chains with a free end.

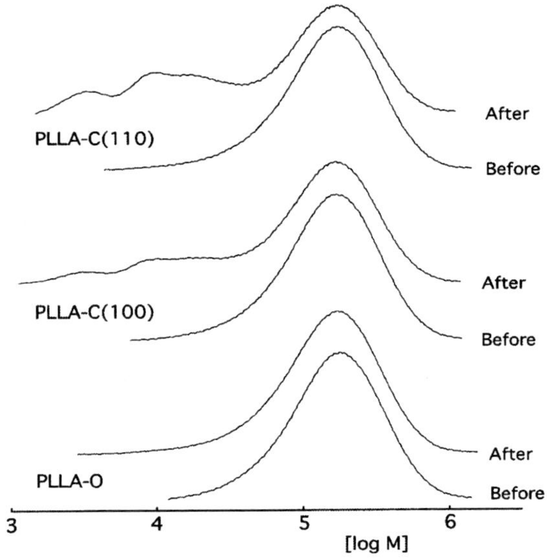

Figure 21. Molecular weight distribution curves of biaxially oriented and de-oriented PLLA films before degradation and after proteinase K-catalyzed enzymatic degradation for 30 hours. PLLA-O is biaxially oriented PLLA film, whereas deoriented PLLA-C(100) and PLLA-C(110) films were prepared by melting PLLA-O at 200°C and crystallized at 100 and 110°C [46].

It is interesting to note that the biaxially oriented PLLA-O film showed no such specific peaks, despite the fact that PLLA-O film had a crystalline thickness similar to that of the de-oriented PLLA-C(110) film, which was prepared by crystallization at 110°C from the melt. This means that the crystalline residues of PLLA-O film were released from the film surface during enzymatic degradation, leaving a negligibly small amount of the crystalline residues on the film surface. Facile release of the crystalline residues may have arisen from the fact that the crystalline regions of PLLA-O film were oriented with their c-axis parallel to the film surface, in marked contrast to the random orientation of the crystalline regions in PLLA-C films.

3.2.3. Effects of Polymer Blending and Pore Formation

Tsuji and Miyauchi also revealed that the incorporation of PDLA or PDLLA reduces proteinase K-catalyzed enzymatic degradation rate of PLLA compared with that expected from the degradation rates of pure PLLA, PDLA, and PDLLA [83]. In other words, the incorporation of PDLA or PDLLA disturbs the proteinase K-catalyzed enzymatic degradation of PLLA. It seems that incorporation of a polymer miscible with PLLA, such as PDLA, PDLLA, and P(LLA-CL) reduces the proteinase K-catalyzed enzymatic degradation [83,85,101]. In contrast, incorporation of other polymers such as poly(vinyl alcohol) or PCL, which are immiscible with PLLA, enhances the proteinase K-catalyzed enzymatic degradation of PLLA (Figure 22) [54,81]. In Figure 22, fraction of PLLA (X_{PLLA}) in PLLA/PCL blends is defined by the following equation:

$$X_{PLLA} = W_{PLLA} / (W_{PLLA} + W_{PCL}) \tag{10}$$

where W_{PLLA} and W_{PCL} are respectively weights of PLLA and PCL in the blend. In these materials, PLLA and the second polymer are phase-separated to form their domains, and enzyme molecules can diffuse into the interfacial area between PLLA domains and the domains of the second polymer (Figure 23) [102,103]. As a result, the degradable surface area of PLLA domains becomes larger than that in pure PLLA materials and, thereby, the enzymatic degradation rate of the blends increases. Reversely, the enzymatic biodegradation rate can be used as an index for the miscibility of the polymer blends containing PLLA. A higher enzymatic biodegradation rate of PLLA in a polymer blend compared to that of pure PLLA indicates that the polymer blend is phase-separated, whereas a lower enzymatic degradation rate of PLLA in a polymer blend compared to that of pure PLLA means that the polymer components are miscible with each other.

On the other hand, the pore formation of biodegradable materials has the same effect as the addition of an immiscible second polymer, and can enhance the enzymatic degradation of biodegradable materials [104]. The pore formation methods are summarized in review articles [3,11]. The removal of the yellow-colored second polymer domains in Figure 23 reveals the enzymatic degradation model of a porous PLLA material. The aforementioned selective enzymatic degradation of phase-separated biodegradable polyester blends can be utilized for the preparation of porous biodegradable materials [81] and for visualization of the phase structure [105]. Here, one component is attacked by the enzyme, selectively degraded, and removed, whereas another component remains unattacked.

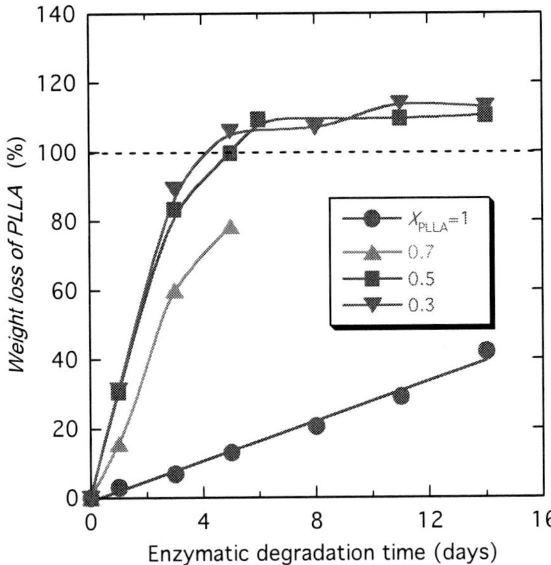

Figure 22. Weight loss of PLLA in PLLA/PCL blends with different X_{PLLA} values during proteinase K-catalyzed enzymatic degradation [81].

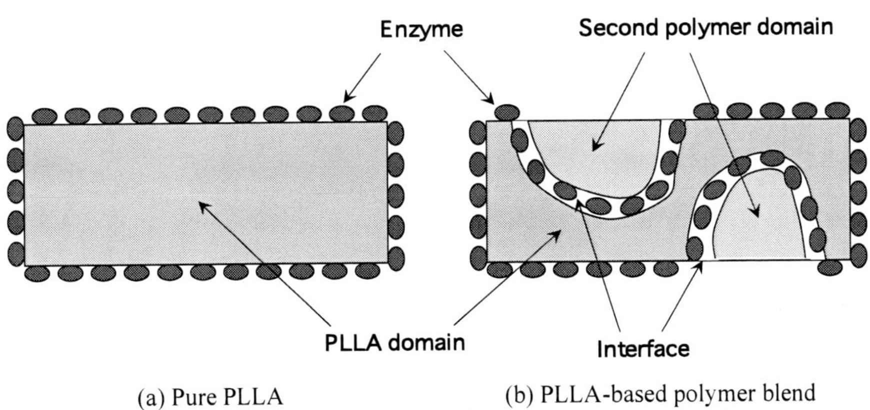

Figure 23. Schematic representation of enzymatic degradation of pure PLLA (a) and PLLA-based polymer blend (b) [102,103].

The combination of two or more methods can be applied to adjust the required biodegradability and mechanical proformance. One example is the combination of pore formation and polymer blending [93]. Porous blends from PLLA/PCL were prepared by water-extraction of poly(ethylene glycol) (PEG) from phase-separated PLLA/PCL/PEG blends. Interestingly, the proteinase K-catalyzed enzymatic degradation of PLLA was extremely enhanced by the combination of the two methods. Also, we developed a method to control the biodegradability of phase-separated PLLA/PCL blends by varying the share rate and time for melt-blending, thereby altering the domain sizes of the two phases [94].

3.2.4. Effects of Surface Treatment

There are two advantages to utilizing the surface treatments of biodegradable polyesters, such as coating and alkaline treatment. That is, with surface treatments one can manipulate biodegradability without altering the bulk physical properties of the materials, such as mechanical properties, and the treatments are applicable on the spot after the production of materials. Among the surface treatments, coating with a biodegradation-resistant polymer is an effective method for reducing biodegradability. In such materials, the surface of biodegradable polymer is protected by the biodegradation-resistant polymer. Tsuji and Ishida indicated that PVA coating is effective in protecting PLLA films from proteinase K and the protective effect is higher with PVA concentration in a coating solution [86].

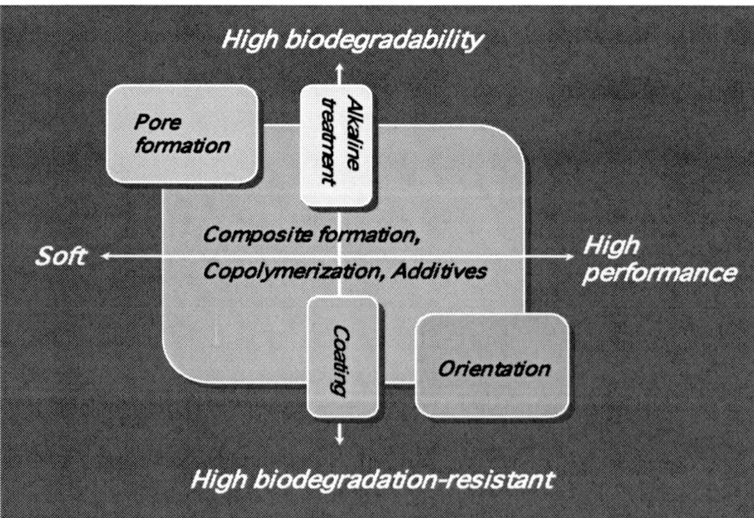

Figure 24. The effects of preparation methods on the biodegradation and mechanical performance of PLA-based materials [102,103].

In contrast, alkaline treatment is effective in enhancing the hydrophilicity of PLA-based materials [20,87], but not so effective for the enhancement of proteinase K-catalyzed enzymatic degradation, compared with the result for lipase-catalyzed enzymatic degradation of PCL [87]. Finally, the effects of various methods on the biodegradation (or enzymatic degradation) and the mechanical performance of PLA-based materials are summarized in Figure 24 [102,103].

3.3. ENVIRONMENTAL AND MICROBIAL DEGRADATION

As shown in Table 5 [61,62,106-141], environmental or microbial degradation of PLA-based materials has been intensively investigated in soil, compost, seawater, or with environmental microbes. PLA-based materials were found to be relatively resistant to biodegradation. Figure 25 shows SEM images of PLLA after biodegradation in soil for 20 months [123]. Most of the surface retained its initial state, except for a small part of the surface where enzymes from filamentous fungi formed grooves with a width of about 3 μm. Figure 26 shows the weight remaining and molecular weight of PLLA/PCL blends having different X_{PLLA} values during biodegradation in soil. The degradation rates of PLLA traced by gravimetry and gel permeation chromatography (GPC) were much lower than those of PCL, and the degradation rates decrease with increasing X_{PLLA}. These findings indicate that the PLLA-degrading microbes are rare or their number per unit volume is very low compared with that of PCL-degrading microbes. It seems that the degradation of PLLA proceeded mostly via abiotic hydrolytic degradation. Based on the conditions that the initial weight-average molecular weight (M_w) of PLLA is 1.33×10^6 g mol^{-1}, the degradation takes place exponentially, and the degradation rate is not influenced by the molecular weight change, the period of time required for PLLA to be degraded to lactic acid is calculated to be 508 months (42.3 years). However, from the viewpoint that CO_2, which was originally in the air, is fixed in PLLA, such a low degradation rate means that CO_2 is retained by PLLA for a long period in the environment. Although it may reduce the circulation rate of carbon in the environment, it will reduce the increasing rate of CO_2 in the air.

The microbes reported to degrade PLLA or PDLLA are given in Table 5. Torres et al. investigated the environmental degradation mechanism of PDLLA in the soil or by microbes from the soil [139]. PDLLA is assumed to be first degraded by chemical hydrolytic degradation, followed by bioassimilation of degraded low-molecular-weight lactic acid oligomers and monomers.

Table 5. Degradation of PLA-based materials in the environment or by microbes [61,62,106-141]

PLA or Copolymer	Microbe or Environment	Enzyme type	Researcher [Reference]
PLLA	Amycolatopsis sp. strain HT 32, 3118, KT-s-9, 41, K104-1	Protease	Pranamuda et al. [104-108], Ikura and Kudo [109], Tokiwa et al. [110], Nakamura et al. [111], Jarerat et al. [112]
	Amycolatopsis orientalis strain IFO12362	Protease	Jarerat et al. [113]
	Bacillus sinithii strain PL 21	Lipase (Esterase)	Sakai et al. [114]
	Brevibacillus (formerly Bacillus brevis)	Protease	Tomita et al. [115]
	Frametes hirsuta strain 2108	-	Cai et al. [116]
	Fusarium moniliforme / Pseudomonas putida	-	Torres et al. [117]
	Fusarium strain 2138, F037, L017, L021, L023	-	Cai et al. [116]
	Geastrum triples strain 2167	-	Cai et al. [116]
	Geobacillus thermocatenulatus	Protease	Tomita et al. [118]
	Kibdelosporangium aridum	Protease	Jarerat et al. [119]
	Lentzea waywayandensis (formerly Saccharothix waywayandensis)	Protease	Jarerat et al. [120], Jarerat et al. [113]
	Polyporous strain 2154	-	Cai et al. [116]
	Stereum strain 2039	-	Cai et al. [116]
	Tritirachium album strain ATCC 22563	Protease	Jarerat and Tokiwa [121], Jarerat et al. [113].
	Soil	-	Södergard et al. [122], Tsuji et al. [123], Ho et al. [124]
	Seawater	-	Tsuji and Suzuyoshi [125,126]

PLA or Copolymer	Microbe or Environment	Enzyme type	Researcher [Reference]
	Compost or its microbes	-	Buchanan et al. [127], Meinander et al. [128], Karjomaa et al. [129], Ho et al. [130], Hakkarainen et al.[131], Gattin [132], Ghorpade [133], Viljanmaa et al. [134], Kale et al. [135]
PDLA	Bacillus stearothermophilus	Protease	Tomita et al. [136]
PDLLA	Cryptococcus sp. strain S-2	Lipase (Cutinase)	Masaki et al. [137]
	Fusarium moniliforme	-	Torres et al. [138]
	Fusarium moniliforme / Pseudomonas putida (Each and mixed culture)	-	Torres et al. [117,139,140]
	Paenibacillus amylolyticus strain TB-13	Lipase	Shigeno et al. [141]
	Penicillium roqueforti	-	Torres et al. [138]
	Soil	-	Torres et al. [139], Ho et al. [124]
	Compost	-	Ho et al. [130]
P(LLA-GA)	Fusarium strain L023	-	Cai et al. [116]
P(DLLA-GA)	Fusarium moniliform	-	Torres et al. [138]

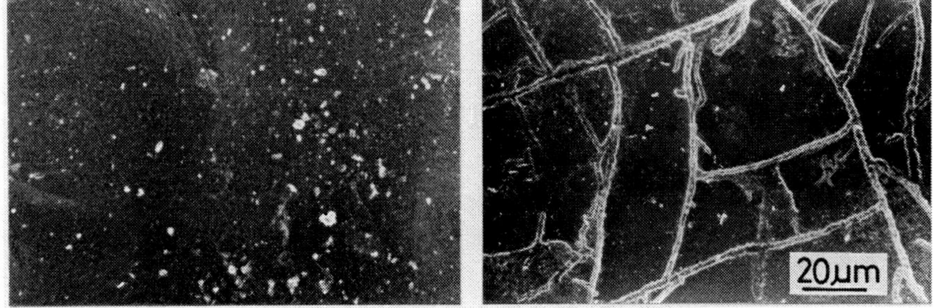

Figure 25. SEM image of PLLA films after biodegradation in soil for 20 months [123].

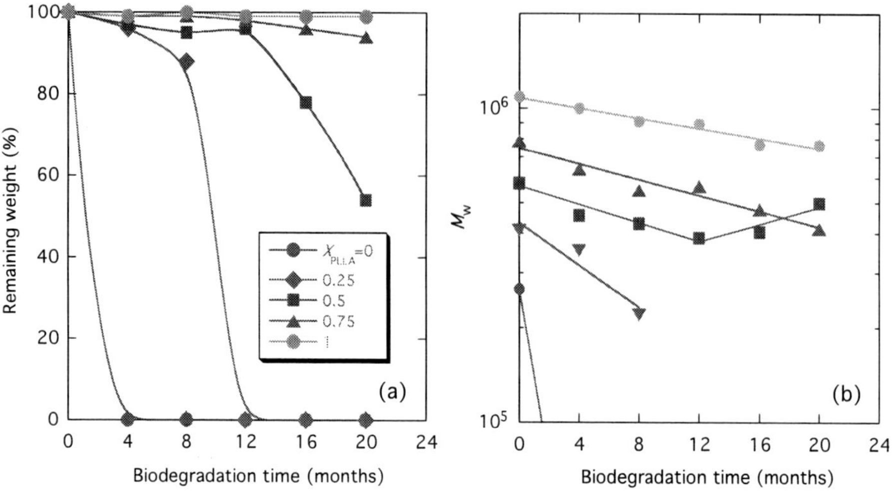

Figure 26. Remaining weight (a) and molecular weight (b) of PLLA/PCL blends with different X_{PLLA} values biodegraded in soil as a function of biodegradation time [123].

The mineralization of PLLA to CO_2 takes place in soils [124] and in compost [127]. Södergard et al. found that peroxide modification slows down the environmental degradation of PLLA in soil [122]. Hakkarainen et al. indicated that the microbes in compost induced the preferred degradation near the chain ends of a PLLA material, and formed lactic acid and lactoyl lactic acid [131]. In this process, ethyl ester of lactoyl lactic acid was formed. Jarerat et al. revealed that the presence of amino acid, peptide, and poly(L-amino acid) enhances the microbial biodegradation of PLLA [113]. Recently, Tomita and coworkers reported that an environmental microbe can degrade PDLA [136]. This indicates that if PLA stereocomplex materials composed of PLLA and PDLA are released into the environment, they will be degraded by environmental microbes.

However, there have been no investigations on the environmental degradation of PLA-based materials in seawater. We have studied the environmental degradation of crystallized and amorphous PLLA (abbreviated as PLLA-C and PLLA-A, respectively) and found that the degradation rate of PLLA in seawater is very low [125]. Figure 27 shows the weight loss of PLLA in seawater, together with those of representative aliphatic polyesters, R-P(3HB) and PCL [125]. Degradation in seawater decreased in the following order: PCL > R-P(3HB) > PLLA. Figure 28 shows polarized photomicrographs of PLLA films before and after exposure to seawater for 10 weeks [125].

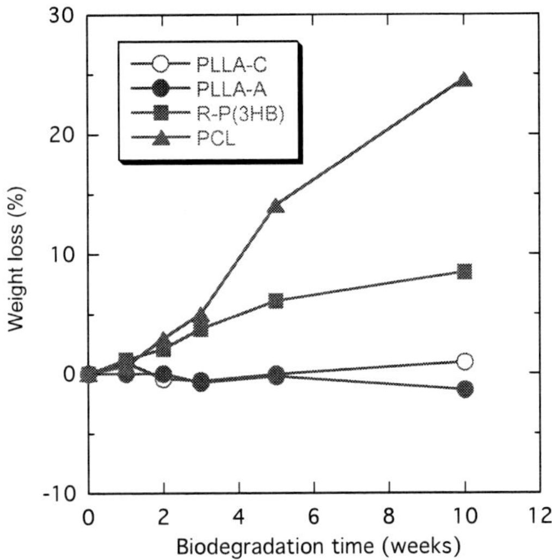

Figure 27. Weight loss of crystallized PLLA (PLLA-C), amorphous PLLA (PLLA-A), R-P(3HB), and PCL exposed to seawater at 25°C [125].

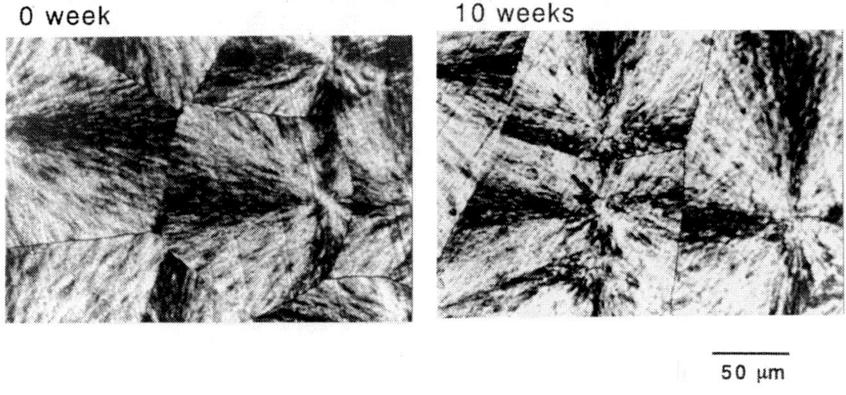

Figure 28. Polarized photomicrographs of PLLA films before and after exposure to seawater for 10 weeks [125].

Although some microbes may have become attached to the PLLA film surface, PLLA films remained unchanged after exposure to seawater. This is in marked contrast to R-P(3HB) and PCL films after biodegradation, where numerous pores were formed by sea microbes (photos not shown here) [125]. Moreover, the stress or strain in seawater from waves and currents caused mechanical destruction of the biodegradable aliphatic polyesters [126]. Such an effect was intense for glassy materials having high T_g such as PLLA.

The environmental or microbial degradation of biodegradable polyesters can be enhanced by pore formation and alkaline surface treatment [142,143]. Pore formation increases the degradable surface per unit mass, and thereby accelerates biodegradation in seawater [143], similar to enzymatic degradation [104]. During surface treatment with an alkaline solution, alkalis induce the cleavage of ester groups only on the film surface, resulting in the formation of carboxyl and hydroxyl groups. This increases the surface hydrophilicity of biodegradable polyesters. Such hydrophilic surfaces attract microbes and elevate the biodegradation rate.

Chapter 5

THERMAL DEGRADATION

Upon the heating of PLA over 250°C, volatile components, such as lactide (LA), cyclic oligomers from trimers to hexamers, CO, CO_2 acetaldehyde, and methylketene, are formed [144]. Figure 29 shows the probable nonradical reactions of PLA thermal degradation [144]. LA and cyclic oligomers are formed by intramolecular transesterification reactions, (a) and (b). An intermolecular transesterification reaction (c) is anticipated not to alter the overall molecular weight distribution of PLA, whereas chain cleavages by a hydrolytic degradation reaction (d) and cis-elimination reaction (e) reduce the molecular weight. The cis-elimination reaction (e) is the same reaction as the Norrish II type photodegradation reaction, resulting in the formation of double bonds at the chain terminals.

The thermal stability and degradation rate of PLA and the formation rate, total yield, and enantiomeric fractions of LA depend on (1) the molecular structures, (2) the type and concentration of the catalyst, (3) the concentration of LA and water in PLA, (4) the type and pressure of the surrounding gas, the method, and other reaction conditions such as temperature and time [145-151]. Table 6 shows the fractions of LLA, DLA, and meso-lactide (MLA) formed by the thermal degradation (depolymerization) of PLLA films [weight average molecular weight (M_w)=4.4x10^5 g mol^{-1}, M_w/number average molecular weight (M_n) =2.1, Sn concentration was about 3-4 ppm] at different temperatures and times [148]. As seen in Table 6, the yield of LA increased with degradation time, irrespective of degradation temperature, and became saturated to around 12% for the degradation periods above 10 hours. The initial formation rate and the saturation yield of LA increased with degradation temperature. The higher yield of LA at higher degradation temperature implies that the reactions of LA formation from PLLA are endothermic [152]. The highest yield of LA (14%) was obtained at 270°C and 10 hours. This value is much lower than that obtained by a small scale distillation system (89%) [147], in which formed LA was removed by evaporation from a reactor.

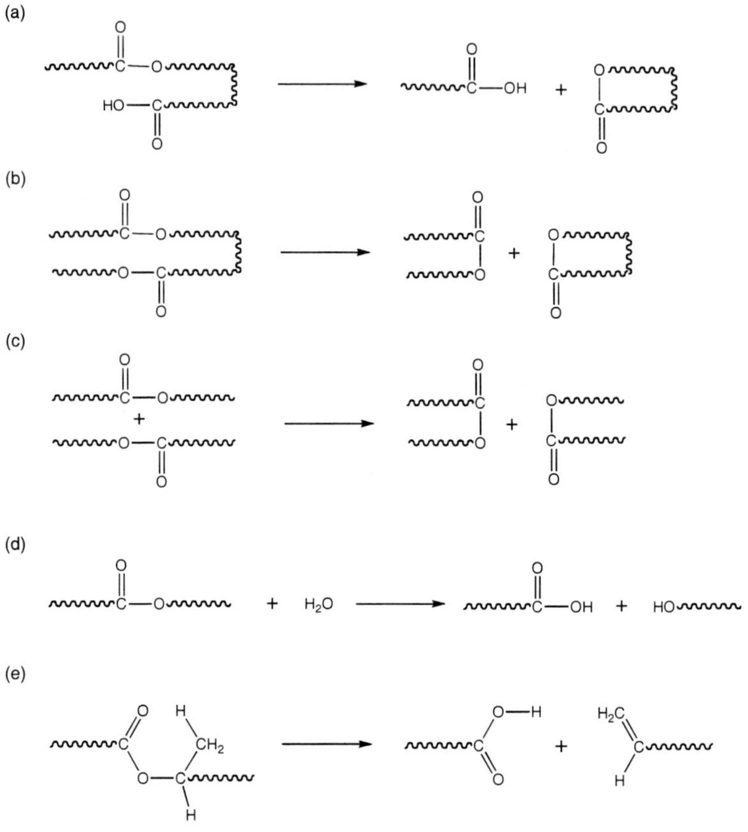

Figure 29. Probable nonradical reactions of PLA thermal degradation: (a) intramolecular transesterification from the end of the chain (back-biting); (b) intramolecular transesterification in the middle of the chain; (c) intermolecular transesterification; (d) hydrolytic degradation; (e) cis- elimination [144].

On the other hand, similar to the yield of LA, the yield of LLA increased with degradation time within 10 hours, irrespective of degradation temperature, and became saturated to around 6% for the periods above 10 hours. The obtained values were much lower than the 88% reported for the small-scale distillation system [147].

Table 6. Yields of LA and LLA, and fractions of LLA, DLA, and MLA formed as a result of thermal degradation of PLLA films at different temperatures and times [148]

Degradation conditions		Yield of LA (%)	Yield of LLA (%)	Fractions of LA		
Temperature (°C)	Time (h)			LLA (%)	DLA (%)	MLA (%)
250	5	3.9	3.6	90.6	1.0	8.4
	10	10.0	8.0	79.5	3.1	17.4
	15	9.4	5.4	57.2	10.1	32.7
270	5	8.9	5.9	66.4	6.9	26.7
	10	14.0	7.3	52.0	12.6	35.4
	15	10.6	4.3	40.7	19.8	39.5
290	5	10.8	5.6	52.1	27.0	20.9
	10	11.9	4.9	42.4	24.3	33.3
	15	12.9	6.1	47.2	20.9	31.9

In a large-scale industrial plant for recycling PLLA to LA, most parts of PLLA are exposed to the conditions like this closed system. Therefore, the large-scale industrial plant has numerous issues to be solved, but thermal depolymerization of PLLA by the small-scale distillation system is effective in obtaining LLA at high yield. That is, an effective recovery method and procedures for LLA formed in a highly viscous PLLA melt should be developed. As seen in Table 6, the highest yield of LLA (8%) was obtained at 250°C and 10 hours. Besides the low yield of LA, the low yield of LLA is attributable to the formation of high amounts of MLA and DLA, especially when the degradation temperature increases. For the formation of MLA and DLA, configurational inversion at the asymmetric carbon atoms in the PLLA chains should occur. In a nonradical thermal degradation process, however, no configurational change at the asymmetric carbon atoms takes place [153,154]. The inversion of configuration at the asymmetric carbon atoms is expected to occur through the formation of carbon atoms having an sp^2 orbital by radical homolysis [Figure 30(a) and(b)] and by enolization [Figure 30(c)] [148]. These pathways were proposed by McNeill and Leiper [153] and by Kopinke and Mackenzie [154], respectively. McNeill and Leiper revealed the significant contribution of such radical homolysis pathways, even at 230°C [153].

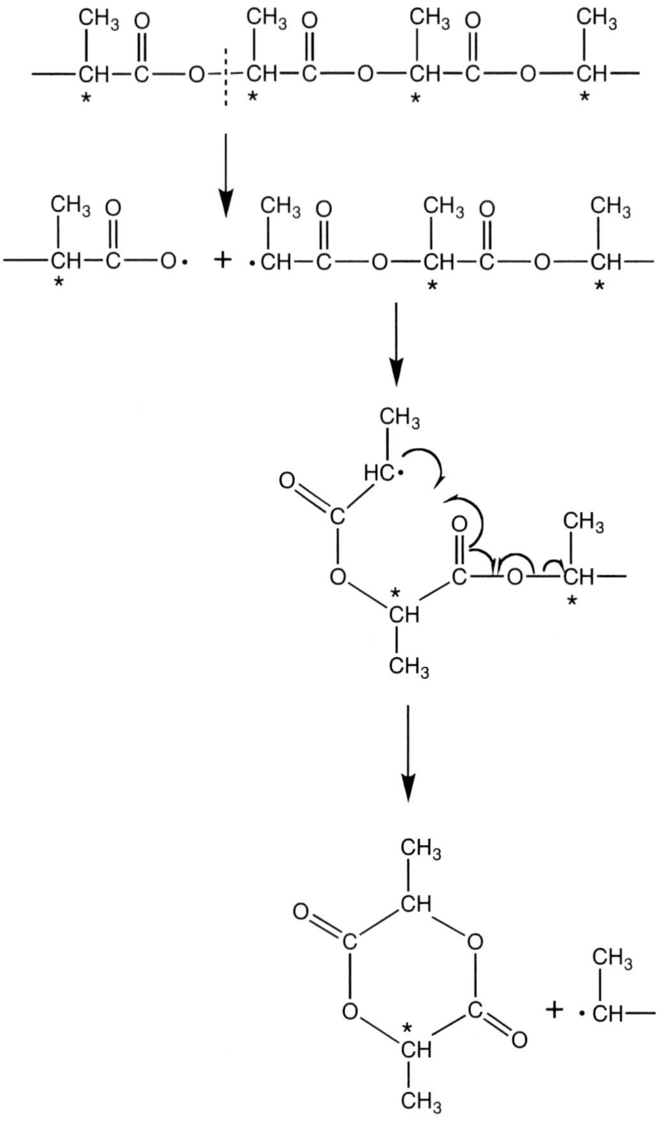

(a)   Alkil-oxygen radical-homolysis and LA formation.

Figure 30. Probable pathways for racemization during LA formation via radical pathways at asymmetric carbon atoms in PLLA: (a) Alkyl-oxygen radical-homolysis and LA formation; (b) Acyl-oxygen radical-homolysis and LA formation; (c) Enolization [148].

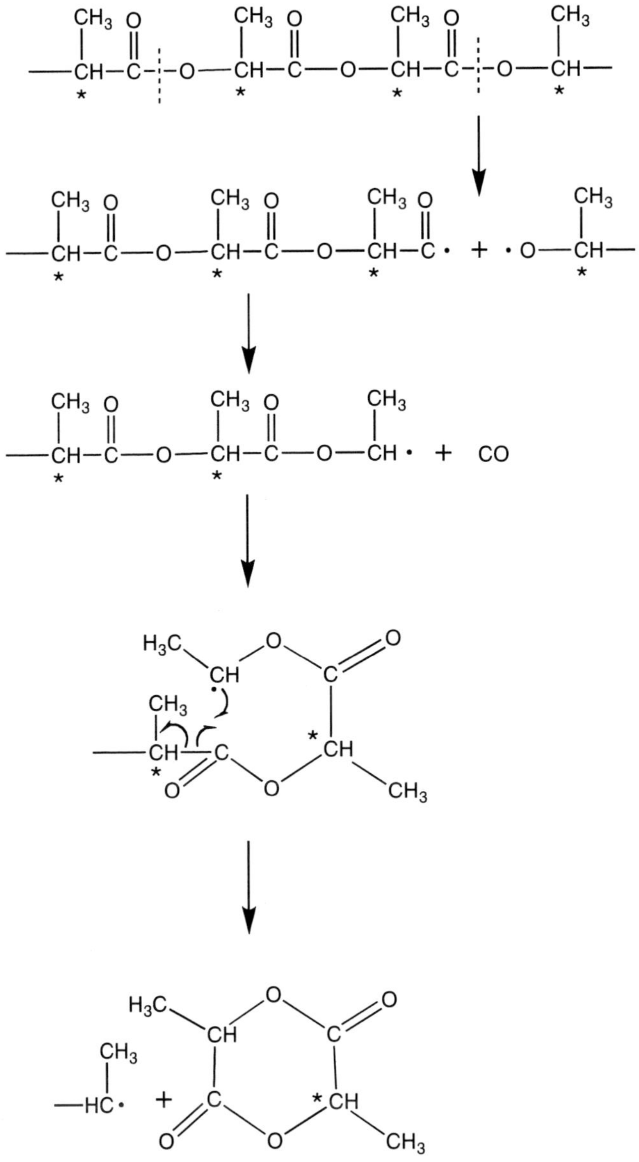

(b) Acyl-oxygen radical-homolysis and LA formation.

Figure 30. Continued

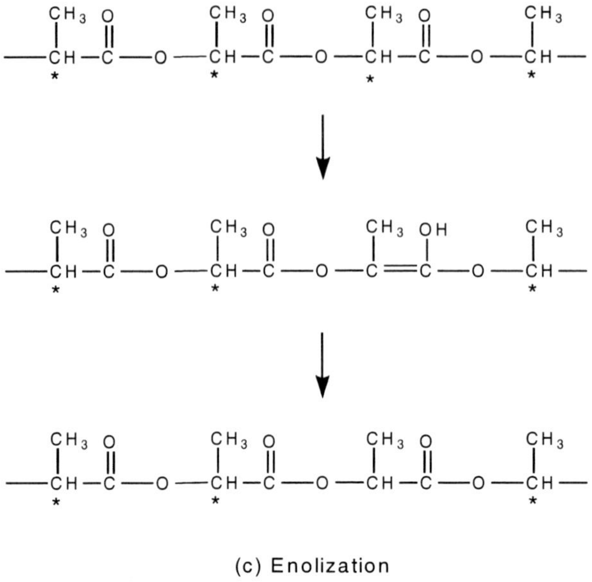

(c) Enolization

Figure 30. Continued

In Figure 30, the asymmetric carbon atoms with an asterisk retain the initial L-configuration, while those without an asterisk can have either a D- or L-configuration. Cyclization at the chain terminals after the radical homolysis can give MLA as well as LLA [Figure 30(a) and (b)]. It should be noted that the terminal carbon atoms having an sp^2 orbital are reproduced after LA formation [Figure 30(a) and (b)], which can form another MLA by cyclization. After configurational inversion by enolization [Figure 30(c)], both MLA and DLA can be formed even if nonradical LA formation takes place. However, for the formation of DLA, a configuration of two sequential asymmetric carbon atoms should be inverted by double enolization, or by single enolization and the subsequent radical homolysis at the neighboring carbon atom, and vice versa. The lower probability for configurational inversion at two sequential carbon atoms compared with that at only one carbon atom must be the cause for the lower fraction of DLA compared to that of MLA, excluding the result at 290°C and 5 min (Table 6). Moreover, the increased fractions of MLA and DLA in LA at a higher temperature strongly suggest that the contribution of the radical homolysis pathways and/or the enolization pathway increased with increasing the degradation temperature.

As stated above, the hydrolytic degradation-resistance of PLA-based materials was enhanced by stereocomplex formation. However, even in the molten state (i.e., above the melting temperature of stereocomplex crystallites), the PLLA/PDLA blend has a higher thermal-resistance compared to that of pure PLLA or PDLA [149]. Figure 31 shows the residual weight of a PLLA/PDLA blend and pure PLLA and PDLA as a function of thermal degradation time at 250 and 260°C [149]. As shown, the PLLA/PDLA blend had a higher residual weight, when compared at the same degradation times.

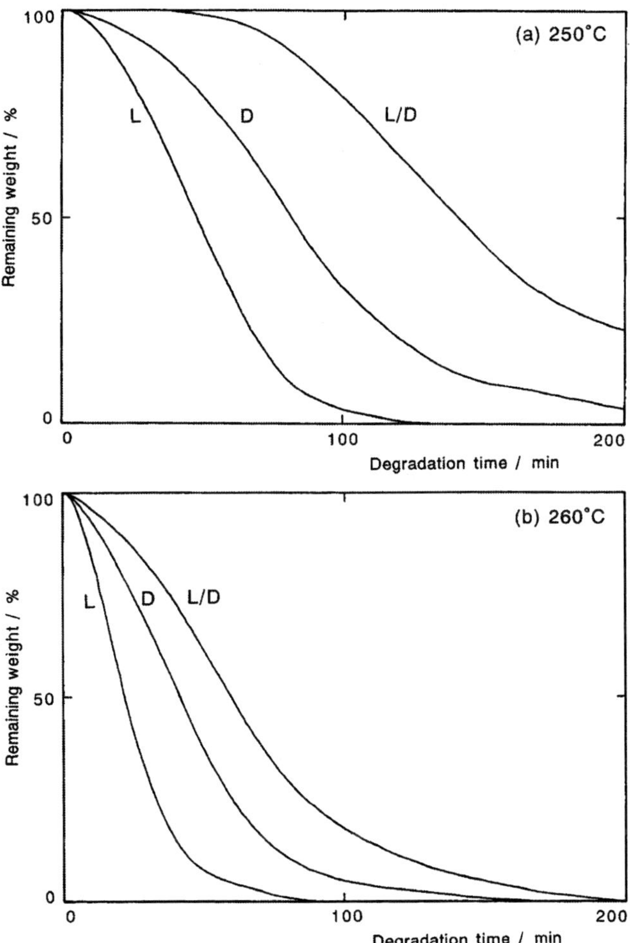

Figure 31. The percentage of remaining weight of the films measured isothermally at (a) 250°C and (b) 260°C as a function of degradation time [149].

Chapter 6

PHOTODEGRADATION

Photodegradation can be the main degradation route of PLA-based materials, when used outdoors or released into the environment after use. However, there have been few reports on the photodegradation of PLA-based materials. Ikada suggested that the Norrish II mechanism is associated with the ultraviolet (UV) light-induced chain cleavage of PLA and PCL [155]. The Norrish II mechanism of PLLA and PCL is shown Figure 32 [156].

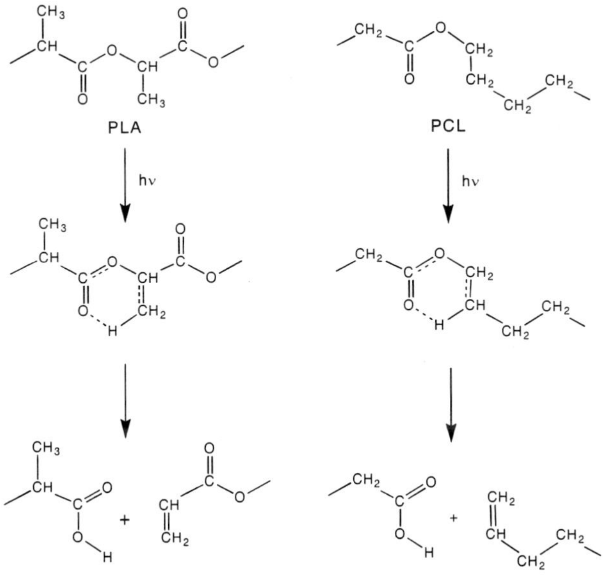

Figure 32. Photodegradation of PLA and PCL via Norrish II mechanism [156].

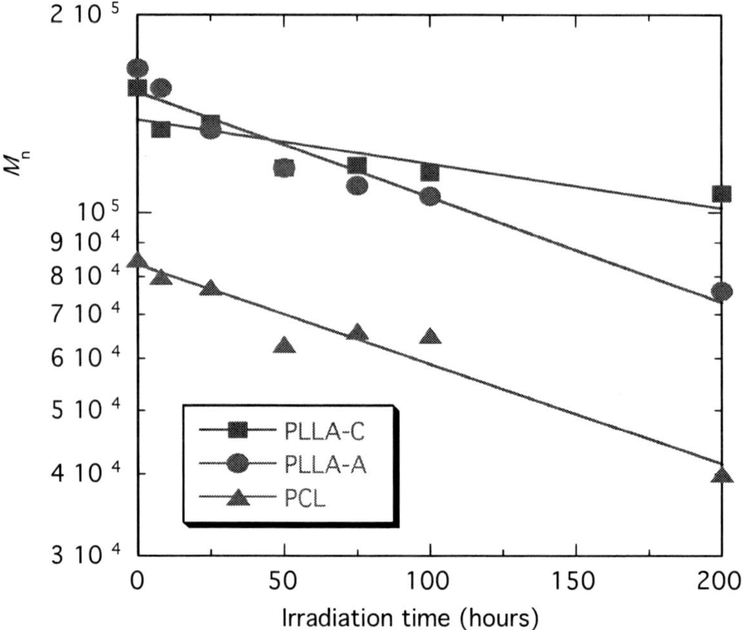

Figure 33. M_n of crystallized PLLA (PLLA-C), amorphous PLLA (PLLA-A), and PCL films as a function of UV-irradiation time [156].

From this figure, PLLA has a higher number of ester groups per unit mass than that of PCL and, therefore, is expected to have a higher photodegradation rate. Figure 33 shows the M_n change of crystallized PLLA (PLLA-C, X_c=34.6%), amorphous PLLA (PLLA-A, X_c=1.0%), and PCL (X_c=56.9%) films with respect to photodegradation carried out by ISO 4892-4 (Plastics-Method of exposure to laboratory light sources-Part 4: Open flame carbon-arc lamps). The photodegradation rate constant (k) can be calculated from Figure 33 according to equation (1). The obtained k values were 1.56×10^{-3}, 3.69×10^{-3}, and 3.50×10^{-3} h^{-1} for the PLLA-C, PLLA-A, and PCL films, respectively.

The k values of PLLA increased with an increase in X_c. This indicates that the chains in the crystalline regions are photodegradation-resistant compared with those in the amorphous region. Furthermore, despite the high number of ester groups per unit mass and the lower X_c value of the PLLA-C film compared with those of the PCL film, the k value of the PLLA-C film is lower than that of the PCL film. This reveals that the number of ester groups per unit mass is not the dominant factor in determining the photodegradation rate, but the molecular structure neighboring on the ester group is a crucial factor.

It is also probable that another mechanism other than the Norrish II type plays an important role in the photodegradation of biodegradable polyesters. Ikada also suggested that the photodegradability of aliphatic polymers is closely related to their biodegradability [155].

Also, we have carried out accelerated photodegradation of PLLA in the presence of *N,N,N',N'*-tetramethyl-1,4-phenylenediamine (TMPD) as a photosensitizer [157]. Accelerated photodegradation experiments revealed that the chain in the crystalline regions can be cleaved by UV-irradiation, which is in marked contrast to the chain cleavability in hydrolytic degradation. Figure 34 shows the molecular weight distribution of PLLA-A and PLLA-C before and after UV-irradiation for 60 hours [157].

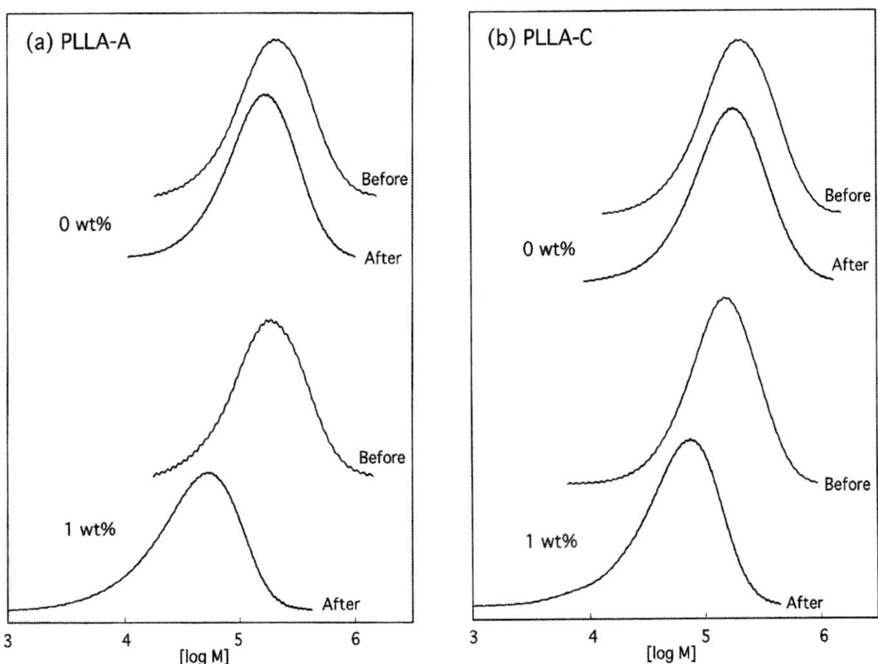

Figure 34. Molecular weight distribution of PLLA-A and PLLA-C with and without photosensitizer before and after UV-irradiation for 60 hours [157].

Copinet found that the molecular weight decrease of PLLA during hydrolytic degradation is accelerated by simultaneous irradiation of UV light [158]. On the other hand, Ho et al. studied the effects of sunlight and rain on the degradation of PLLA and PDLLA in banana fields [159]. We investigated the proteinase K-catalyzed degradation of UV-irradiated (60 hours) and non-irradiated PLLA specimens, and found that enzymatic degradation at an early stage was enhanced by UV-irradiation, although disturbed at a late stage [160]. The acceleration at an early stage is due to the lowered molecular weight of PLLA caused by the chain cleavage of UV-irradiation, whereas the disturbed degradation at a late stage should be ascribed to the accumulation of double bonds and/or crosslinks formed by UV-irradiation.

Chapter 7

CONCLUSION

The rate of hydrolytic degradation and biodegradation can be manipulated by varying molecular and highly ordered structures, fillers, material shape, and surface structure. Recently, PLA-based materials are utilized in industrial and commodity applications as well as biomedical and pharmaceutical applications. To widen the applications of PLA-based materials, higher mechanical performance, thermal stability, hydrolytic degradation- and photodegradation-resistance are required, in addition to the reduction of the production cost.

REFERENCES

[1] Kharas G.B., Sanchez-Riera F., and Severson D.K. (1994). *Polymer of lactic acids.* In Mobley D.P. (Ed.). *Plastics from microbes* (pp. 93-137). New York: Hanser Publishers.

[2] Doi Y., and Fukuda K. editors. (1994). *Biodegradable plastics and polymers.* Amsterdam: Elsevier.

[3] Coombes A.G.A., and Meikle M.C. (1994). Bioabsorbable synthetic polymers as replacements for bone graft, *Clin. Mater. 17*, 35-67.

[4] Vert M., Schwarch G., Coudane J.J. (1995). Present and future of PLA polymers. *J. Macromol. Sci.: Pure Appl. Chem.* 1995; A32: 787-796.

[5] Hartmann, M.H. (1998). High molecular weight polylactic acid polymers. In Kaplan, D.L., (Eds.), *Biopolymers from renewable resources* (pp. 367-411). Berlin (Germany): Springer; 1998.

[6] Ikada, Y., and Tsuji, H. (2000). Biodegradable polyesters for medical and ecological applications. *Macromol. Rapid Commun. 21*, 117-132.

[7] Garlotta, D. (2001). A literature review of poly(lactic acid). *J. Polym. Environ., 9*, 63-84.

[8] Albertsson, A.-C. (Ed.) (2002). *Degradable aliphatic polyesters (Advances in polymer science, vol.157).* Berlin (Germany): Springer.

[9] Södergård, A., and Stolt, M. (2002). Properties of lactic acid based polymers and their correlation with composition. *Prog. Polym. Sci. 27*, 1123-1163.

[10] Scott, G. (Ed.) (2002). *Biodegradable polymers: Principles and applications* (2nd edition). Dordrecht (The Netherlands): Kluwer Academic Publishers; 2002.

[11]   Tsuji, H. (2002). Poly(lactide). In Doi, Y. and Steinbüchel, A. (Eds.), *Polyesters III (Biopolymers, vol. 4)* (pp. 129-177). Weinheim (Germany): Wiley-VCH.

[12]   Auras, R., Harte, B., and Selke, S. (2004). An overview of polylactides as packaging materials. *Macromol. Biosci., 4*, 835-864.

[13]   Tsuji, H. (2005). Poly(lactide) Stereocomplexes: Formation, structure, properties, degradation, and applications. *Macromol. Biosci.*, 2005, 5, 569-597.

[14]   Tsuji, H. (2002). *Biodegradable polymers* (in Japanese). Tokyo (Japan): Corona Publishing Co., Ltd.

[15]   von Burkersroda, F., Schedl, L., and Göpferich, A. (2002). Why degradable polymers undergo surface erosion or bulk erosion. *Biomaterials, 23*, 4221-4231.

[16]   Li, S., and Vert, M. (1995). Biodedegradation of aliphatic polyesters. In Scott, G., andGilead, D. (Eds.), *Biodegradable polymers: Principles and applications* (pp. 43-87). Cambridge: Chapman and Hall.

[17]   Makino, K., Arakawa, M., and Kondo, T. (1985). Preparation and in vitro degradation properties of polylactide microcapsules. *Chem. Pharm. Bull. 33*, 1195-1201.

[18]   Cam, D., Hyon, S.-H., and Ikada, Y. (1995). Degradation of high molecular weight poly(L-lactide) in alkaline medium. *Biomaterials, 16*, 833-843.

[19]   Tsuji, H., and Ikada, Y. (1998). Properties and morphology of poly(L-lactide): II. Hydrolysis in alkaine solution, *J. Polym, Sci. :Part A: Polym. Chem., 36*, 59-66.

[20]   Tsuji, H., and Ishida, T. (2003). Poly(L-lactide): X. Enhanced surface hydrophilicity and chain-scission mechanisms of poly(L-lactide) film in enzymatic, alkaline, and phosphate-buffered solutions. *J. Appl. Polym. Sci., 87*, 1628-1633.

[21]   Li, S., Girod-Holland, S., and Vert, M. (1996). Hydrolytic degradation of poly(-lactic acid) in the presence of caffeine base. *J. Control. Release, 40*, 41-53.

[22]   Tsuji, H., and Tezuka, Y. (2005). Alkaline and enzymatic degradation of L-lactide copolymers: 1. Amorphous-made films of L-lactide copolymers with D-lactide, glycolide, ε-caprolactone. *Macromol. Biosci., 5*, 135-148.

[23]   Tsuji, H., and Miyauchi, S. (2001). Poly(L-lactide): 7. Enzymatic hydrolysis of free and restricted amorphous regions in poly(L-lactide) films with different cystallinities and a fixed crystalline thickness. *Polymer, 42*, 4463-4467.

[24] Tsuji, H., and Nakahara, K. (2002). Poly(L-lactide): IX. Hydrolysis in acid media. *J. Appl. Polym. Sci.,* 86, 186-194.

[25] Tsuji, H., and Ikarashi, K. (2004). In vitro hydrolysis of poly(L-lactide) crystalline residues as extended-chain crystallites: III. Effects of pH and enzyme. *Polym. Degrad. Stab., 85,* 647-656.

[26] Reed, A.M., and Gilding, D.K. (1981) Biodegradable polymers for use in surgery-Poly(glycolic)/poly(lactic acid) homo and copolymers: 2. In vitro degradation. *Polymer, 22,* 494-498.

[27] Tsuji, H., Daimon, H., and Fujie, K. (2003). A new strategy for recycling and preparation of poly(L-lactic acid): Hydrolysis in the melt. *Biomacromolecules, 4,* 835-840.

[28] Saeki, T., Tsukegi, T., Tsuji, H., Daimon, H., and Fujie, K. (2005). Hydrolytic degradation of poly[(R)-3-hydroxybutyric acid] in the melt. *Polymer, 46,* 2157-2162.

[29] Tsuji, H., Ono, T., Saeki, T., Daimon, H., and Fujie, K. Hydrolytic degradation of poly(ε-caprolactone) in the melt. *Polym. Degrad. Stab., 89,* 336-343.

[30] Campanelli, J.R., Kamal, M.R., and Cooper, D.G. (1993). A kinetic study of the hydrolytic degradation of polyethylene terephthalate at high temperatures. *J. Appl. Polym. Sci. 48,* 443-451.

[31] Tsuji, H., Nakahara, K., and Ikarashi, K. (2001). Poly(L-lactide): 8. High-temperature hydrolysis of poly(L-lactide) films with different crystallinities and crystalline thicknesses in phosphate-buffered solution. *Macromol. Mater. Eng., 286,* 398-406.

[32] Tsuji, H., Ikarashi, K., and Fukuda N. (2004). Poly(L-lactide): XII. Formation, growth, and morphology of crystalline residues as extended-chain crystallites through hydrolysis of poly(L-lactide) films in phosphate-buffered solution. *Polym. Degrad. Stab., 84,* 515-523.

[33] Tsuji, H., and Ikarashi, K. (2004). In vitro hydrolysis of poly(L-lactide) crystalline residues as extended-chain crystallites: II. Effects of hydrolysis temperature. *Biomacromolecules, 5,* 1021-1028.

[34] Saha, S.K., and Tsuji, H. (2006). Hydrolytic degradation of amorphous-made films of L-lactide copolymers with glycolide and D-lactide, *Macromol. Mater. Eng., 291,* 357-368.

[35] Tsuji, H., Mizuno, A., and Ikada, Y. (2000) Properties and morphologies of poly(L-lactide). III. Effects of initial crystallinity on long-term in vitro hydrolysis of high molecular weight poly(L-lactide), *J. Appl. Polym. Sci., 77,* 1452-1464.

[36]   Wu, L., and Ding, J. (2004). In vitro degradation of three-dimensional porous poly(D,L-lactide-*co*-glycolide) scaffolds for tissue engineering. *Biomaterials*, *25*, 5821-5830.

[37]   Malin, M., Hiljanen-Vainio, M., Karjalainen, T., Seppälä, J. (1996). Biodegradable lactone copolymers: II. Hydrolytic study of -caprolactone and lactide copolymers. *J. Appl. Polym. Sci.*, 59, 1289-1298.

[38]   Vert, M., Chabot, F., Leray, P., and Christel, P. (1981). Stereoregular bioresorbable polyesters for orthopaedic surgery. *Makromol. Chem. Suppl.*, 5, 30-41.

[39]   Tsuji, H. (2002). Autocatalytic hydrolysis of amorphous-made polylactides: Effects of L-lactide content, tacticity, and enantiomeric polymer blending, *Polymer, 43*, 1789-1796.

[40]   Saha, S.K., and Tsuji, H. (2006). Effects of molecular weight and small amounts of D-lactide units on hydrolytic degradation of poly(L-lactic acid)s. *Polym. Degrad. Stab., 91*, 1665-1673.

[41]   Hyon, S.-H., Jamshidi, K., and Ikada, Y. (1998). Effects of residual monomer on the degradation of DL-lactide polymer. *Polym. Int., 46*, 196-202.

[42]   Nakamura, T., Hitomi, S., Watanabe, S., Shimizu, Y., Jamshidi, K., Hyon, S.-H., and Ikada, Y. (1989). Bioabsorption of polylactides with different molecular properties, *J. Biomed. Mater. Res., 23,* 1115-1130.

[43]   Li, S., Garreau, H., and Vert, M. (1990). Structure-property Relationships in the case of the degradation of massive poly(α-hydroxy acids) in aqueous media: Part 3. Influence of the morphology of poly(L-lactic acid). *J. Mater. Sci., Mater. Med., 1*, 198-206.

[44]   Duek, E.A.R., Zavaglia, C.A.C., and Belangero, W.D. (1999). In vitro study of poly(lactic acid) pin degradation. *Polymer, 40*, 6465-6473.

[45]   Cai, H. Dave, V. Gross, R.A., and McCarthy, S.P. (1996). Effects of physical aging, crystallinity, and orientation on the enzymatic degradation of poly(lactic acid). *J. Polym. Sci.: Part B: Polym. Phys., 34*, 2701-2708.

[46]   Tsuji, H., Ogiwara, M., Saha, S.K., and Sakaki, T. (2006). Enzymatic, alkaline, and autocatalytic degradation of poly(L-lactic acid): Effects of biaxial orientation, *Biomacromolecules, 7,* 380-387.

[47]   Hyon, S.-H., Jamshidi, K., and Ikada, Y. (1984). Melt spinning of poly-L-lactide and hydrolysis of the fiber in vitro. In Shalaby, S.W., Hoffman, A.S., Ratner, B.D., and Horbett, T.A. (Eds.), *Polymers as Biomaterials* (pp. 51-65). NewYork: Prenum Press.

[48] Hoogsteen, W., Postema, A.R., Pennings, A.J., ten Brinke, G., and Zugenmaier, P. (1990). Crystal structure, donformation, and morphology of solution-spun poly(L-lactide) fibers, *Macromolecules, 23*, 634-642.

[49] Okihara, T., Tsuji, M., Kawaguchi, A., Katayama, K., Tsuji, H., Hyon, S.-H., and Ikada, Y. (1991). Crystal structure of stereocomplex of poly(L-lactide) and poly(D-lactide), *J. Macromol. Sci.-Phys. B30*, 119-140.

[50] For example, Gedde, U.W. (1995). *Polymer physics* (pp.131-198). London: Chapman and Hall.

[51] Kalb, B., and Pennings, A.J. (1980). General crystallization behaviour of poly(L-lactic acid), *Polymer, 21*, 607-612.

[52] Tsuji, H., and Ikada, Y. (1995). Properties and morphologies of poly(L-lactide): 1. annealing condition effects on properties and morphologies of poly(L-lactide), *Polymer, 36*, 2709-2716.

[53] Pitt, C.G., Cha, Y., Shah, S.S., and Zhu, K.J. (1992). Blends of PVA and PGLA: Control of the permeability and degradability of hydrogels by blending, *J. Controlled Release, 19*, 189-200.

[54] Tsuji, H., and Muramatsu, H. (2001). Blends of aliphatic polyesters: 5. Non-enzymatic and enzymatic hydrolysis of blends from hydrophobic poly(L-lactide) and hydrophilic poly(vinyl alcohol), *Polym. Degrad. Stab., 71*, 403-413.

[55] Nijenhuis, A.J., Colstee, E., Grijpma, D.W., Pennings, A.J. (1996). High molecular weight poly(L-lactide) and poly(ethylene oxide) blends: Thermal characterization and physical properties, *Polymer, 37*, 5849-5857.

[56] Kishida, A., Yoshioka, S., Takeda, Y., Uchiyama, M. (1989). Formulation-assisted biodegradable polymer matrices. *Chem. Pharm. Bull., 37*, 1954-1956.

[57] Renouf-Glauser, A.C., Rose, J., Farrar, D., and Cameron, R.E. (2005). A degradation study of PLLA containing lauric acid. *Biomaterials, 26*, 2415-2422.

[58] Tsuji, H., Ikada, Y. (2000). Properties and morphology of poly(L-lactide): 4. Effects of structural parameters on long-term in vitro hydrolysis of poly(L-lactide) in phosphate-buffered solution. *Polym. Degrad. Stab., 67*, 179-189.

[59] Tsuji, H. (2000). In vitro hydrolysis of blends from enantiomeric poly(lactide)s: 1. Well-stereocomplexed blend and non-blended films. *Polymer, 41*, 3621-3630.

[60] Lam, K.H., Nieuwenhuis, P., Molenaar, I., Esselbrugge, H., Feijen, J., Dijkstra, P.J., Shakenraad, J.M. (1994). Biodegradation of porous versus non-porous poly(L-lactic acid) dilms. *J. Mater. Sci. Mater. Med.*, *5*, 181-189.

[61] Faisal, M., Saeki, T., Tsuji, H., Daimon, H., and Fujie, K. (2007). Depolymerization of poly(L-lactic acid) under hydrothermal conditions. *Asian J. Chem.*, *19*, 1714-1722.

[62] Tokiwa, Y., and Jarerat, A. (2004). Biodegradation of poly(l-lactide). *Biotech. Lett.*, *26*, 771-777.

[63] Tokiwa, Y., and Calabia, P. (2006). Biodegradability and biodegradation of poly(lactide). *Appl. Microbiol. Biotech.*, *72*, 244-251.

[64] Williams, D.F. (1981). Enzymatic hydrolysis of polylactic acid. *Eng. Med.*, *10*, 5-7.

[65] Reeve, M.S., McCarthy, S.P., Downey, M.J., and Gross, R.A. (1994). Polylactide stereochemisty: Effect on enzymatic degradability. *Macromolecules, 27*,825-831.

[66] Gajria, A.M., Davé, V., Gross, R.A., and McCarthy, S.P. (1996). Miscibility and biodegradability of blends of poly(lactic acid) and poly(vinyl acetate). *Polymer*, *37*, 437-444.

[67] MacDonald, R.T., McCarthy, S.P., and Gross, R.A. (1996). Enzymatic degradability of poly(lactide): Effects of chain stereochemistry and material crystallinity. *Macromolecules*, *29*, 7356-7361.

[68] Sheth, M., Kumar, R.A., Davé, V., Gross, R.A., and McCarthy, S.P. (1997). Bioodegrdable polymer blends from poly(lactic acid) and poly(ethylene glycol), *J. Appl. Polym. Sci.*, *66*, 1495-1505.

[69] Wang, L., Ma, W. Gross, R. A., and McCarthy, S.P. (1998). Reactive compatibilization of biodegradable blends of poly(lactic acid) and poly(ε-caprolactone). *Polym. Degrad. Stab.*, *59,* 161-168.

[70] Nagata, M. Okano, F. Sakai, and W. Tsutsumi, N. (1998). Separation and enzymatic degradation of blend films of poly(L-lactic acid) and cellulose, *J. Polym. Sci.: Part A: Polym. Chem.*, *36*, 1861-1864.

[71] Li, S., McCarthy, S. (1999). Influence of crystallinity and stereochemistry on the enzymatic degradation of poly(lacticde)s. *Macromolecules*, *32*, 4454-4456.

[72] Liu, L., Li, S., Garreau, and H., Vert, M. (2000). Selective enzymatic degradations of poly(L-lactide) and poly(ε-caprolactone) blend films. *Biomacromolecules*, *1*, 350-359.

[73] Li, S., Tenon, M., Garreau, H., Braud, C., and Vert, M. (2000). Enzymatic degradation of stereocomplymers derived from L-, DL- and meso-lactide. *Polym. Degrad. Stab. 67*, 85-90.

[74] Li, S., Girard, A., Garreau, H., and Vert, M. (2001). Enzymatic degradation of polylactide stereocompolymers with predominant D-lactyl content. *Polym. Degrad. Stab., 71*, 61-67.

[75] Shirahama, H., Mizuma, K., Umemoto, K., and Yasuda, H. (2001). Synthesis of poly(lactide-ran-MOHEL) and its biodegradation with proteinase K. *J. Polym. Sci.: Part A: Polym. Chem., 39*, 1374-1381.

[76] Shirahama, H., Tanaka, A., and Yasuda, H. (2002). Highly biodegradable copolymers composed of chiral depsipeptide and L-lactide units with favorable physical properties. *J. Polym. Sci.: Part A: Polym. Chem. 40*, 302-316.

[77] Tsutsumi, C., Shirahama, H., and Yasuda, H. (2002). Enzymatic degradations of copolymers of L-lactide with cyclic carbonates. *Macromol. Biosci., 2*, 223-232.

[78] Tsutsumi, C., Nakagawa, K., Shirahama, H., and Yasuda, H. (2003). Biodegradation of statistical copolymers composed of D,L-lactide and cyclic carbonate. *Polym. Int., 52*, 439-447.

[79] Watanabe, Y., Shirahama, H., and Yasuda, H. (2004). Synthesis of random copolymers lactides with 1,5-dioxepan-2one and their biodegadability. *React. Funct. Polym., 59*, 211-224.

[80] Shirahama, H., Ichimaru, A. Tsutsumi, C., Nakayama, Y., and Yasuda, H. (2005). Characteristics of the biodegradability and physical properties of stereocomplexes between Poly(L-lactide) and Poly(D-lactide) Copolymers, *J. Polym. Sci., Part A: Polym,. Chem., 43*, 438-454.

[81] Tsuji, H., and Ishizaka T. (2001). Porous biodegradable polyesters: 3. Preparation of porous poly(ε-caprolactone) films from blends by selective enzymatic removal of poly(L-lactide). *Macromol. Biosci. 1*, 59-65.

[82] Tsuji, H., and Miyauchi, S. (2001). Poly(L-lactide): 6. Effects of crystallinity on enzymatic hydrolysis of poly(L-lactide) without free amorphous region. *Polym. Degrad. Stab., 71*, 415-424.

[83] Tsuji, H., and Miyauchi, S. (2001). Enzymatic hydrolysis of poly(lactide)s: Effects of molecular weight, L-lactide content, and enantiomeric and diastereoisomeric polymer blending. *Biomacromolecules, 2*, 597-604.

[84] Fukuda, N., Tsuji, H., and Ohnishi Y. (2002). Physical properties and enzymatic hydrolysis of poly(L-lactide)-$CaCO_3$ composites, *Polym. Degrad. Stab., 78*, 119-127.

[85]   Tsuji, H., and Yamada. T. (2003). Blends of aliphatic polyesters: VIII. Effects of poly(L-lactide-co-ε-caprolactone) on enzymatic hydrolysis of poly(L-lactide), poly(ε-caprolactone), and their blend films. *J. Appl. Polym. Sci., 87,* 412-419.

[86]   Tsuji, H., and Ishida, T. (2003). Surface hydrophilicity and enzymatic hydrolyzability of biodegradable polyesters: 2. Effects of hydrophilic polymer coating; *Macromol. Biosci., 3,* 51-59.

[87]   Tsuji, H., Ishida, T., and Fukuda, N. (2003). Surface hydrophilicity and enzymatic hydrolyzability of biodegradable polyesters: 1. Effects of alkaline treatment. *Polym. Int., 52,* 843-852.

[88]   Fukuda, N. and Tsuji, H. (2005). Physical properties and enzymatic hydrolysis of poly(L-lactide)-TiO$_2$ composites. *J. Appl. Polym. Sci., 96,* 190-199.

[89]   Tsuji, H., Tezuka, Y., and Yamada, K. (2005). Alkaline and enzymatic degradation of L-lactide copolymers: 2. Crystallized films of poly(L-lactide-co-D-lactide) and poly(L-lactide) with similar crystallinities, *J. Polym. Sci., Part B: Polym. Phys., 49,* 1064-1075.

[90]   Tsuji, H., Echizen, Y., and Nishimura Y. (2006). Enzymatic degradation of poly(L-lactic acid): Effects of UV irradiation, *J. Polym. Environ., 14,* 239-248.

[91]   Tsuji, H., Kidokoro, Y., and Mochizuki, M. (2006). Enzymatic degradation of biodegradable polyester composites of poly(L-lactic acid) and poly(ε-caprolactone). *Macromol. Mater. Eng., 291,* 1245-1254.

[92]   Tsuji, H., Kidokoro, Y., and Mochizuki, M. (2007). Enzymatic degradation of poly(L-lactic acid) fibers: Effects of small drawing, *J. Appl. Polym. Sci., 103,* 2064-2071.

[93]   Tsuji, H., and Horikawa, G. Porous biodegradable polyester blends of poly(L-lactic acid) and poly(ε-caprolactone): Physical properties, morphology, and biodegradation. *Polym. Int., 56,* 258-266.

[94]   Tsuji, H., Horikawa, and G., Itsuno, S. (2007). Melt-processed biodegradable polyester blends of poly(L-lactic acid) and poly(ε-caprolactone): Effects of processing conditions on biodegradation. *J. Appl. Polym. Sci., 104,* 831-841.

[95]   Yamashita, K., Kikkawa, Y., Kurokawa, K., andDoi, Y. (2005). Enzymatic degradation of poly(L-lactide) film by proteinase K: Quartz crystal microbalance and atomic force microscopy study. *Biomacromolecules, 6,* 850-857.

[96]   Fukuzaki, H., Yoshida, M., Asano, M., and Kumakura, M. (1989). Synthesis of copoly(-lactic acid) with relatively low molecular weight and *in vitro* degradation. *Eur. Polym. J.*, *25*, 1019-1026.

[97]   Ivanova, T.Z., Panaiotov, I., Boury, F., Proust, J.E., and Verger, R. (1997). Enzymatic hydrolysis of poly(D,L-lactide) spread monolayers by cutinase. *Colloid Polym. Sci.*, *275*, 449-457.

[98]   Oda, Y., Yonetsu, A., Urakami, T., and Tonomura, K. (2000). Degradation of Polylactide by Commercial Proteases. *J. Polym. Environ.*, *8*, 29-32.

[99]   Iwata, T. andDoi, Y (1998). Morphology and enzymatic degradation of poly(L-lactic acid) single crystals. *Macromolecules*, *31*, 2461-2467.

[100]   Kikkawa, Y., Abe, H., Iwata, T., Inoue, Y., and Doi, Y. (2002). Crystallization, stability and enzymatic degradation of poly(L-lactide) thin film. *Biomacromolecules*, *3*, 350-356.

[101]   Tsuji, H., Yamada, T., Suzuki, M., and Itsuno, S. (2003). Blends of aliphatic polyesters: VII. Effects of poly(L-lactide-co-ε-caprolactone) on morphology, structure, crystallization, and physical properties of blends of poly(L-lactide) and poly(ε-caprolactone). *Polym. Int.*, *52*, 269-275.

[102]   Tsuji, H. (2007), Degradation mechanisms and development of biodegradable polymers (in Japanese). In Mechanisms and analyses of biodegradation of organic substances (in press). Tokyo: Johoukikou, Co., Ltd.

[103]   Tsuji, H., *Poly(lactic acid)s* (in Japanese), to be published, Chiba (Japan): Komeda Publishing Co., Ltd.

[104]   Tsuji, H., and Ishizaka T. (2001). Porous biodegradable polyesters: 2.Physical properties, morphology, and enzymatic and alkaline hydrolysis of porous poly(ε-caprolactone) films. *J. Appl. Polym. Sci.*, *80*, 2281-2291.

[105]   Tsuji, H. and Ishizaka, T. (2001). Blends of aliphatic polyesters: VI. Lipase-catalyzed hydrolysis and visualized phase structure of biodegradable blends from poly(ε-caprolactone) and poly(L-lactide). *Int. J. Biol. Macromol.*, *29*, 83-89.

[106]   Pranamuda, H. Tokiwa, Y., and Tanaka, H. (1997) Polylactide degradation by an *Amycolatopsis* sp. *Appl. Environ. Microbiol.*, *63*, 1637-1640.

[107]   Pranamuda, H. and Tokiwa, Y. (1999) Degradation of poly(L-lactide) by strains belonging to *Genus. amycolatopsis*, *Biotech. Lett.*, *21*, 901-905.

[108]   Pranamuda, H., Tsuchii, A., and Tokiwa, Y. (2001). Poly(l-lactide)-degrading enzyme produced by Amycolatopsis sp. *Macromol. Biosci.* 1, 25–29.

[109] Ikura, Y., and Kudo, T. (1999). Isolation of a microorganism capable of degrading poly(l-lactide). *J. Gen. Appl. Microbiol., 45*, 247–251.

[110] Tokiwa, Y., Konno, M., and Nishida, H. (1999). Isolation of silk degrading microorganisms and its poly(L-lactide) degradability, *Chem. Lett.*, 355-356.

[111] Nakamura, K., Tomita, T., Abe, N., and Kamio, Y. (2001). Purification and characterization of an extracellular poly(l-lactic acid) depolymerase from a soil isolate, *Amycolatopsis* sp. strain K104-1. *Appl. Environ. Microbiol. 67*, 345–353.

[112] Jarerat, A., Pranamuda, H., and Tokiwa, Y. (2002). Poly(l-lactide)-degrading activity in various *Actinomycetes*. *Macromol. Biosci., 2*, 420–428.

[113] Jarerat, A., Tokiwa, Y., and Tanaka, H. (2004). Microbial poly(l-lactide)-degrading enzyme induced by amino acids, peptides and poly(l-amino acids). *J. Polym. Environ. 12*, 139–146.

[114] Sakai, K., Kawano, H., Iwami, A., Nakamura, M., and Moriguchi, M. (2001). Isolation of a thermophilic poly-l-lactide degrading bacterium from compost and its enzymatic Characterization. *J. Biosci. Bioeng. 92*, 298–300.

[115] Tomita, K., Kuroki, Y., and Nagai, K. (1999). Isolation of thermophiles degrading poly(l-lactic acid). *J. Biosci. Bioeng., 87*,752–755.

[116] Cai, Q., Bei, J., Luo, A., and Wang, S. (2001). Biodegradation behaviour of poly(lactide-co-glycolide) induced by microorganisms, *Polym. Degrad. Stab. 71*, 243-251.

[117] Torres, A., Li, S.M., Roussos, S., and Vert, M. (1996). Degradation of L- and DL-lactic aid oligomers in the presence of *Fusarium Moniliforme* and *Pseudomonas Putida, J. Environ. Polym. Degrad., 4*, 213-223.

[118] Tomita, K., Nakajima, T., Kikuchi, Y., and Miwa, N. (2004). Degradation of poly(l-lactic acid) by a newly isolated thermophile. *Polym. Degrad. Stab. 84*, 433–438.

[119] Jarerat, A., and Tokiwa, Y. (2003). Poly(l-lactide) degradation by Kibdelosporangium aridum. *Biotechnol. Lett. 25*, 2035–2038.

[120] Jarerat, A., and Tokiwa, Y. (2003). Poly(l-lactide) degradation by Saccharotrix waywayandensis. *Biotechnol. Lett., 25*,401–404.

[121] Jarerat, A., and Tokiwa, Y. (2001). Degradation of poly(l-lactide) by Fungus. *Macromol. Biosci., 1*, 136–140.

[122] Södergård, A., Selin, J.-F., and Pantke, M. (1996). Environmental degradation of Peroxide modified poly(L-lactide), *Int. Biodeter. Biodegrad., 38*, 101-106.

[123] Tsuji, H., Mizuno, A., and Ikada, Y. (1998). Blends of aliphatic polyesters: III. biodegradation of solution-cast blends from poly(L-lactide) and poly(ε-caprolactone), *J. Appl, Polym, Sci.*, *70*, 2259-2268.

[124] Ho, K.-L.G., and Pometto, A.L. III. (1999). Temperature effects on soil mineralization of polylactic acid plastic in laboratory repspirometers. *J. Environ. Polym. Degrad.*, *7*, 101-108.

[125] Tsuji, H. and K. Suzuyoshi, K. (2002). Environmental degradation of biodegradable polyesters: 1. Poly(ε-caprolactone), poly[(R)-3-hydroxybutyrate], and poly(L-lactide) films in controlled static seawater; *Polym. Degrad. Stab.*, *75*, 347-355 (2002).

[126] Tsuji, H. and K. Suzuyoshi, K. (2002). Environmental degradation of biodegradable polyesters: 2. Poly(ε-caprolactone), poly[(R)-3-hydroxybutyrate], and poly(L-lactide) films in natural dynamic seawater. *Polym. Degrad. Stab.*, *75*, 357-363.

[127] Buchanan, C.M., Dorschel, D.D., Gardner, R.M., Komarek, R.J. and White, A.W. (1995). Biodegradation of cellulose esters: Composting of cellulose ester-diluent mixture. *J. Macromol. Sci., -Pure Appl. Chem.*, *A32*, 683-697.

[128] Meinander, K., Niemi, M., Hakola, J.S., and Selin, J.-F. (1997). Polylactides-degradable polymers for fibers and films, *Macromol. Symp.*, *123*, 147-153.

[129] Karjomaa, S., Suortti, T., Lempiäinen, R., Selin, J.-F., and Itävaara, M. (1998). Microbial degradation of poly-(L-lactic acid) oligomers, *Polym. Degrad. Stab.*, *59*, 333-336.

[130] Ho, K.-L.G., Pometto, A.L. III, Gadea-Rivas A., Briceno, J.A., and Rojas, A. (1999). Degradation of Polylactic Acid (PLA) Plastic in Costa Rican Soil and Iowa State University Compost Rows. *J. Environ. Polym. Degrad.*, *7*, 173-177.

[131] Hakkarainen, M., Karlsson, S.. and Albertsson, A.-C. (2000). Rapid (bio)degradation of polylactide by mixed culture of compost microorganisms-low molecular weight products and matrix changes, *Polymer, 41*, 2331-2338.

[132] Gattin, R., Copinet, A., Bertrand, C., and Couturier, Y. (2001). Comparative Biodegradation Study of Starch- and Polylactic Acid-Based Materials. *J. Polym. Environ, 9*, 11-17.

[133] Ghorpade, V.M., Gennadios, A., and Hanna, M.A.(2001). Laboratory composting of extruded poly(lactic acid) sheets. *Bioresource Technol., 76*, 57-61.

[134] Viljanmaa, M. Södergård, A., Mattila, R., and Törmälä, P. (2002). Hydrolytic and environmental degradation of lactic acid based hot melt adhesives. *Polym. Degrad. Stab., 78*, 269-278.

[135] Kale, G., Auras, R., and Singh S.P. (2007). Comparison of the degradability of poly(lactide) packages in composting and ambient exposure conditions. *Packing Tech. Sci., 20*, 49-70.

[136] Tomita, K., Tsuji, H., Nakajima, T., Kikuchi, Y., Ikarashi, K., Ikeda, N. (2003). Degradation of poly(d-lactic acid) by a thermophile. *Polym. Degrad. Stab. 81*, 167–171.

[137] Masaki, K., Kamini, N.R., Ikeda, H., Iefuji, H. (2005). Cutinase-Like Enzyme from the Yeast *Cryptococcus* sp. Strain S-2 Hydrolyzes Polylactic Acid and Other Biodegradable Plastics. *Appl. Environ Microbiol. 71*, 7548-7550.

[138] Torres, A., Li, S.M., Roussos, S., and Vert, M. (1996). Screening of microorganisms of biodegradation of poly(lactic acid) and lactic acid-containing polymers, *Appl. Environ. Microbiol., 62*, 2393-2397.

[139] Torres, A., Li, S.M., Roussos, S., and Vert, M. (1996). Poly(lactic acid) degradation in soil or under controlled conditions. *J. Appl. Polym. Sci., 62*, 2295-2302.

[140] Torres, A., Li, S.M., Roussos, S., and Vert, M. (1999). Microbial degradation of a poly(lactic acid) as a model of synthetic polymer degradation mechanisms in out door conditions, *A.C.S. Symp. Ser., 723*, 218-226.

[141] Shigeno, Y.A., Teeraphatpornchai, T., Teamtisong, K., Nomura, N., Uchiyama, H., Nakahara, T., and Kambe, T.N. (2003). Cloning and sequencing of a poly(dl-lactic acid) depolymerase gene from *Paenibacillus amylolyticus* strain TB-13 and its functional expression in *Escherichia coli. Appl. Environ. Microbiol. 69*, 2498–2504.

[142] Tsuji, H., Suzuyoshi, K., Tezuka, Y., and Ishida T. (2003). Environmental degradation of biodegradable polyesters: 3. Effects of alkali-treatment on biodegradation of poly(ε-caprolactone) and poly[(R)-3-hydroxybutyrate] films in controlled soil. *J. Polym. Environ., 11*, 57-65.

[143] Tsuji, H., and Suzuyoshi, K. (2003). Environmental degradation of biodegradable polyesters: 4. Effects of pores and surface hydrophilicity on biodegradation of poly(ε-caprolactone) and poly[(R)-3-hydroxybutyrate] films in controlled seawater. *J. Appl. Polym. Sci., 90*, 587-593.

[144] Kopinke, F.-D., Remmler, M., Mackenzie, K., Möder, M., and Wachen, O. (1996). Thermal decomposition of biodegradable polyester-II. Poly(lactic acid), *Polym. Degrad. Stab., 53*, 329-342.

[145] Jamshidi, K., Hyon, S.-H., and Ikada, Y. (1995). Thermal characterization of polylactides, *Polymer, 29, 2229-2234.*

[146] Cam, D. and Marucci, M. (1997). Influence of residual monomers and metals on poly (L-lactide) thermal stability, *Polymer, 38,* 1879-1884.

[147] Noda, M., and Okuyama, H. (1999). Thermal catalytic depolymerization of poly(L-lactic aicd) oligomer into LL-lactide: Effects of Al, Ti, Zn and Zr compounds as catalysts. *Chem. Pharm. Bull., 47,* 467-471.

[148] Tsuji, H., Fukui, I., Daimon, H., and Fujie, K. (2003). Poly(L-lactide): XI. Lactide formation by thermal depolymerization of poly(L-lactide) in a closed system. *Polym. Degrad. Stab., 81,* 501-509 (2003).

[149] Tsuji, H., and Fukui, I. (2003). Enhanced thermal stability of poly(lactide)s in the melt by enantiomeric polymer blending. *Polymer, 44,* 2891-2896.

[150] Mori, T., Nishida, H., Shirai, Y., and Endo, T. (2004). Effects of chain end structures on pyrolysis of poly(-lactic acid) containing tin atoms. *Polym. Degrad. Stab., 84, 243-251.*

[151] Fan, Y., Nishida, H., Mori, T., Shirai, Y., and Endo, T. (2004). Thermal degradation of poly(-lactide): effect of alkali earth metal oxides for selective L,L-lactide formation, *Polymer, 45,* 1197-1205.

[152] See for example, Atkins, P.W. (1998). Physical chemistry (sixth edition, Chapter 9, pp. 224-227). Oxford: Oxford Univ. Press.

[153] McNeill, I.C., and Leiper, H.A. (1985). Degradation studies of some polyesters and polycarbonates: 2. Polylactide: Degradation under isothermal conditions, thermal degradation mechanism and photolysis of the polymer. *Polym. Degrad. Stab., 11,* 309-326.

[154] Kopinke, F.-D., and Mackenzie K.J. (1997). Mechanistic aspects of the thermal degradation of poly(lactic acid) and poly(β-hydroxybutyric acid). *Anal. Appl. Pyrolysis, 40-41,* 43-53.

[155] Ikada, E. (1997). Relationship between photodegradability and biodegradability of some aliphatic polyesters. *J. Photopolym. Sci. Tech., 10,* 265-270.

[156] Tsuji, H., Echizen, Y., and Nishimura, Y. (2006). Photodegradation of biodegradable polyesters: A comprehensive study on poly(L-lactide) and poly(ε-caprolactone), *Polym. Degrad. Stab., 91,* 1128-1137.

[157] Tsuji, H., Echizen, Y., Saha, S. K., and Nishimura, Y. (2005). Photodegradation of poly(L-lactic acid): Effects of photosensitizer, *Macromol. Mater. Eng., 290,* 1192-1203.

[158] Copinet, A., Bertrand, C., Govindin, S., Coma, V., and Couturier, Y. (2004). Effects of ultraviolet light (315 nm), temperature and relative humidity on the degradation of polylactic acid plastic films. *Chemosphere, 55*, 763-773.

[159] Ho, K.-L.G., Pometto III, A.L. Hinz, P.N. Gadea-Rivas, A., Briceño, J.A., and Rojas, A. (1999). Field exposure study of polylactic acid (PLA) plastic films in the banana fields of Costa Rica. *J. Polym. Environ., 7,* 167-172.

[160] Tsuji, H., Echizen, Y., Nishimura, Y. (2006). Enzymatic degradation of poly(L-lactic acid): Effects of UV irradiation. *J. Polym. Environ., 14,* 239-248

INDEX

automobile, 1

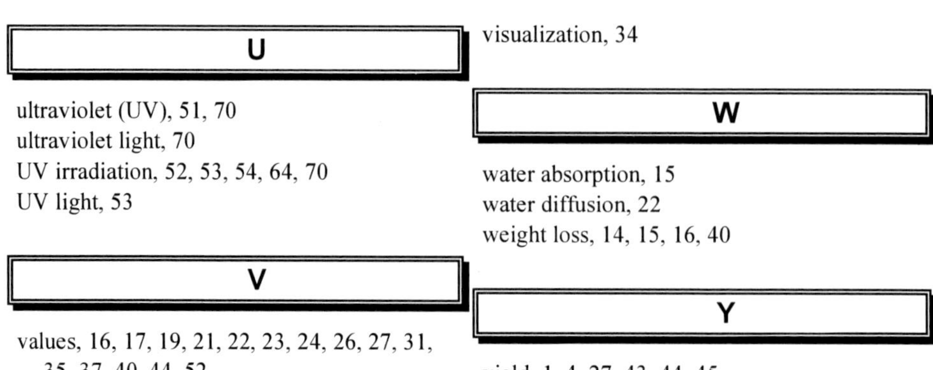